Modeling Chemical Transport in Soils

Natural and Applied Contaminants

Modeling Chemical Transport in Soils

Natural and Applied Contaminants

Edited by

Hossein Ghadiri
Calvin W. Rose
Centre for Catchment and In-stream Research
Division of Environmental Sciences,
Griffith University,
Brisbane, Queensland, Australia

LEWIS PUBLISHERS
Boca Raton Ann Arbor London Tokyo

Library of Congress Cataloging-in-Publication Data

Modeling chemical transport in soils : natural and applied contaminants /
edited by Hossein Ghadiri and Calvin W. Rose
p. cm.
Includes bibliographical references and index.
ISBN 0-87371-747-3
1. Water, Underground—Pollution—Mathematical models.
2. Groundwater flow—Mathematical models. I. Ghadiri, Hossein.
II. Rose, Calvin W.
TD426.M62 1992
628.5'5—dc20

92-16369
CIP

Direct all inquiries to CRC Press, Inc., 2000 Corporate Blvd., N.W.,
Boca Raton, Florida 33431.

PRINTED IN THE UNITED STATES OF AMERICA
1 2 3 4 5 6 7 8 9 0
Printed on acid-free paper

Preface

The growing global concern over soil erosion and the urgent need to develop reliable and effective methods of predicting or estimating soil and chemical transport from agricultural and nonagricultural watersheds into waterways and reservoirs has focused attention on the development of mathematical models of these processes. A large number of models have been developed during the last two or three decades, and modeling these processes is currently pursued in various parts of the world. While the bulk of past modeling work has been carried out in the U.S., scientists from other continents have also contributed significantly to the rapid progress in modeling nonpoint source pollution.

A basic knowledge of the wide range of nonpoint source models developed in various parts of the world, and an understanding of their strengths and weaknesses, is needed by anyone intending to use them, to improve them, or to attempt developing new models.

The aim of this review is to bring together a brief account of the most widely used models, or those with the potential of becoming planning or research tools in nonpoint source pollution, and to critically evaluate them. Transport of sorbed and water-soluble chemicals by overland flow and by subsurface flow have all been covered. Modeling chemical reaction during subsurface flow is also reviewed. General characteristics of nonpoint source models, criteria for choosing a model, the application, and (where possible) comparative evaluation of models has also received attention.

The first two chapters of this book, which deal with sorbed chemical transport modeling and the general introduction to models and modeling processes, have been prepared by the two of us, Hossein Ghadiri and Calvin Rose, who also acted as the joint editors of the book. Chapter 3 on subsurface transport of pollutant and chemical reactions has been written by Dr. D.A. Barry of the Centre for Water Research, the University of Western Australia. Our colleagues from Queensland Department of Primary Industries, Mr. Roger Shaw, Mr. Ian Gordon, and Mr. Peter Thorburn, jointly prepared Chapter 4 on salt movement. Dr. David Charls-Edward of the Centre for Catchment and In-Stream Research of Griffith University kindly acted as our literary editor of the first two chapters. Dr. P. J. Hensley provided useful references to the relevant literature for Chapter 4. To all of these people and the reviewers of this book we are extremely grateful for their helpful assistance.

Hossein Ghadiri
Calvin W. Rose
Brisbane

Acknowledgment

This work has been supported by the Australian Water Research Advisory Committee (AWRAC) now replaced by the Land and Water Resources Research Development Corporation (LWRRDC), Queensland Department of Primary Industries (QDPI) and the Water Authority of Western Australia.

Hossein Ghadiri graduated from Jundi-Shapur University of Iran in 1968 with a B.Sc. degree in Agriculture. He was then awarded a scholarship to study in England where he received his M.Agr.Sc. and Ph.D. degrees in Soil Science from Reading University in 1975 and 1978, respectively. Before joining the Division of Environmental Sciences of Griffith University in 1987, he was an associate professor of soil physics-soil conservation at Jundi-Shapur University in Iran for several years. His broad interests are in soil erosion, raindrop impact and the breakdown of soil aggregates, sediment and sorbed chemical transport by overland flow, and chemical enrichment mechanisms. He is currently involved in pesticide transport characterization and modeling studies conducted at Griffith University.

Calvin W. Rose is the Dean and Professor of Environmental Sciences Division of Griffith University, Australia. He has worked in the aircraft industry, the Commonwealth Scientific and Industrial Research Organization (CSIRO) and Universities in East Africa, the U.S., and Australia. Experience of relevance to this book is in modeling soil erosion and salinity processes, and in investigating the role of soil, water, and chemical processes and environmental constraints in productive systems. He is currently involved in the direction of projects concerned with the management of soil erosion for sustained crop production (supported by ACIAR), and with characteristics and field-to-stream transport of pesticides and other agricultural chemicals (supported by LWRRDC of the Commonwealth Government).

Contributors

D. A. Barry is a graduate of the University of Queensland and Griffith University. Following his doctoral studies, he went to the Virginia Polytechnic Institute and State University in 1985 to perform research on fluid flow and solute transport in heterogeneous domains. He then went to the University of California, Riverside to investigate the application of stochastic differential equation methods to chemical transport in randomly heterogeneous porous media. Since 1989 he has held an academic position in hydrology at the University of Western Australia. His interests include infiltration, nonlinear sorption, multicomponent solute transport, and model verification.

Peter J. Thorburn is a Senior Soil Scientist with the Land Management Research Branch of the Queensland Department of Primary Industries, Australia. He joined the DPI in 1981 to conduct research into the cause and management of soil salinity problems in Queensland. His work focused on the development and application of solute movement models, to provide guidelines for irrigation management in areas with saline irrigation water and shallow water tables. He has also worked extensively in dryland salinity in Queensland. Peter has a B.Agr.Sc. from Queensland University and a M.Phil. degree from Griffith University. He is currently on leave to the Centre for Groundwater Studies, Adelaide (South Australia), were he is conducting research into diffuse discharge of shallow groundwaters.

Roger Shaw is a Principal Soil Scientist with the Queensland Department of Primary Industries. Following graduation in Agricultural Science from the University of Queensland in 1968 he joined the Department. Mr. Shaw has worked on soil water and salinity issues in both dryland and irrigated agriculture. He was awarded a Churchill Fellowship in 1984 for overseas study in irrigation water quality assessment. His expertise is in landscape, soil and water salinity processes and management.

Ian Gordon is a Soil Scientist with the Queensland Department of Primary Industries, working on dryland and irrigated salinity. After receiving a degree in Agricultural Science at the University of Queensland, Mr. Gordon continued with postgraduate studies in mining rehabilitation aiming to develop method for ameliorating saline-sodic mine overburden materials. He joined the Department of Primary Industries in 1989 and has been involved in projects related to salinity and contaminant hydrology.

Table of Contents

CHAPTER 1

An Introduction to Nonpoint Source Pollution Modeling

H. Ghadiri and C. W. Rose

1.1 INTRODUCTION

Water flowing through or over the land surface contains various kinds of dissolved and suspended material. Even rain water is not pure by the time it reaches the ground. While some of these dissolved or suspended materials are of natural origin, some are man-made pollutants or natural material brought to unusually high levels by generated accelerating natural processes. Industrial processes and agricultural activities are important contributors to the lowering of stream-water quality. The growing demand for food and fiber, water, and energy, requires us to understand better the transport of materials and chemicals, and to implement more effective catchment (or watershed) management to maintain acceptable water quality despite changing land use in the catchment. Water quality standards, and the level at which a chemical constituent is regarded as a pollutant, vary with the use to which water is put. For example, quality standards for drinking water are much higher than those necessary for irrigation water. The most common materials that become pollutants when their concentrations exceed certain proscribed levels are described in the following section. The efforts that have been made to model the processes that affect such levels will be discussed in subsequent sections.

The basis for extrapolation of measurements made in a specific context to other geographical areas or conditions can be strengthened by the use of appropriate models. Because of the complex nature of nonpoint source pollution and its extreme

ISBN 0-87371-747-3

variability, the use of statistical methods alone to analyze limited data sets may not be an effective solution to the problem of evaluating nonpoint pollution. Nor are deterministic models, based on average conditions and characteristics, adequate to represent the inherent variabilities. In air-quality modeling it has been found that combined deterministic and stochastic models are mutually helpful. This book reflects the current emphasis in the literature on models of nonpoint pollution that commonly express in mathematical form the physical processes thought to determine the accumulation or deposition, attenuation, and transport of pollutants to the aquatic environment. These processes that affect water quality and hydrologic processes occurring on the land surface or in the soil profile are continuous in nature; hence, continuous forms of models are commonly assumed necessary to their accurate representation. As a corollary, models that simulate single storm events must be put within a realistic and hence somewhat stochastic environmental context if they are to be used to evaluate nonpoint pollution in the longer term.

The representation of nonpoint source pollution by continuous models recognizes that pollution transport mechanisms are universal. Whether the pollutants originate from pervious or impervious lands, from agricultural or urban areas, or from natural or modified lands, the major transport mode of runoff and its associated sediment load is operative. Modeling nonpoint source pollution arising from overland flow has a structure analogous to a three-layered pyramid. The base of the pyramid is the hydrologic behavior of the watershed. Without accurate representation of runoff, modeling nonpoint pollutants in overland flow is impossible. The second layer represents the process of sediment loss, providing the major transport mechanism for all sorbed pollutants. The top layer of the pyramid is the interaction of the various pollutants with sediment loss and runoff that results in the overall transport and delivery of nonpoint source pollutants to streams or other water bodies.

Models can help in forecasting acute problems, such as the short-term impact of pesticide transport, and with the statistical description of the long-term characteristics of chemical transport processes. These latter types of description can help in predicting chronic or long-term impacts resulting from the modification of watersheds. To choose an appropriate model from existing ones, the goals for which the model is to be used and the resources available to meet those goals must be clearly understood and explicitly defined. Models range in complexity from detailed research tools to relatively simple planning tools. A research model may incorporate the best understanding of the processes being modeled, and be constructed to reveal gaps in our knowledge of these processes. Simple planning models may be developed directly, or through the simplification of more detailed models.

1.2 POLLUTING AGENTS

The most common pollutants of water are sediment, nutrients, pesticides, organic matter, heavy metals, salts, microorganisms, and oxygen-demanding material. The first five are briefly described in this chapter and salt will be dealt with in Chapter 4.

1.2.1 Sediment

Sediment, that is, eroded soil, reduces the storage capacities of reservoirs and the carrying capacities of waterways and channels. It also carries, absorbed on the surface of the sediment particles, chemicals, nutrients, pesticides, and heavy metals. Standards for soil loss from croplands have been discussed (Moldenhauer and Onstad, 1975), such as an allowable soil loss of between 3 and 5 tons of soil per acre each year (that is 7 to 14 tons ha^{-1} $year^{-1}$). These were based on the loss of soil resources and not on the subsequent downstream environmental impact of the sediment. Such standards have yet to be developed and a major problem is the complexity of the linkage between eroded sediment and chemical delivery to streams.

The processes of soil erosion and sediment transport by rain and runoff is selective in terms of particle size and weight. In general, the coarser sediments deposit first and the finer ones move further. As a consequence, suspended sediments tend to contain more of the finer and less dense particles. Because these fine particles usually contain higher amounts of adsorbed chemicals it follows that an enrichment of any sorbed chemical in eroded sediment must be considered and included in any prediction of chemical runoff (Young and Onstad, 1976; Rose and Dalal, 1988). Models for predicting chemical enrichment will be discussed later.

1.2.2 Plant Nutrients

Chemicals containing nitrogen and phosphorus are common causes of water-quality problems. These pollutants may lead to human health hazards, but more commonly their loss into groundwater leads to a reduction in soil productivity and eutrophication of receiving waters. Their inorganic forms, such as NO_3^- and NH_4^+ and orthophosphate (PO_4^{3-}), are easily measured. Organic compounds containing nitrogen and phosphorus are numerous and more difficult to measure. Ammonium and orthophosphate are also adsorbed onto mineral particles in the sediment by ion-exchange or sorption reactions. The physical transport of both inorganic and organic forms of nitrogen and phosphorus with sediment is important, and enrichment associated with the processes of deposition and resuspension of sediment must be recognized. The insoluble forms of nitrogen and phosphorus far exceed their soluble forms, and the prediction of the availability of their insoluble forms is difficult. Usually only the total amounts of nitrogen and phosphorus are reported.

Some non-process-based modeling of the inorganic nutrients ammonia, nitrate, nitrite, and orthophosphate has been done. Haith (1976) indicated that 60 to 70% of the variation in nitrate levels could be explained by land management practices, agricultural land use accounting for about 35% of it; however, land use practices did not explain significant variation in phosphorus levels.

Nitrate is quickly leached into the soil profile and not picked up by water runoff unless fertilizer has recently been spread onto the soil surface. A part, usually small, of the nitrate in runoff comes from the rain and another part from infiltrated water that may emerge to the surface again in larger watersheds. Some non-process-based modeling of inorganic nutrients ammonia, nitrate, nitrite, and PO_4^{3-} has been attempted (Haith, 1976). Many models have been developed to predict the behavior of nitrogen and phosphorus in the soil-plant-water system, and some will be discussed in Section 2.5.

A first-order rate equation has been proposed by Smith et al. (1976) to describe nitrogen mineralization. Models of leaching, which removes NO_3^- and NH_4^+ from the root zone, have been developed, and these also have application to the description of salinity and pesticide transport (Rose et al., 1982a,b).

Davidson et al. (1978a) have developed a detailed research model that considers the processes of downward transport of water and dissolved nitrogen, equilibrium adsorption-desorption of ammonium, microbiological transformation of nitrogen, and uptake of water, nitrate, and ammonium. Because this model is very complex, needing a large number of input parameters that are seldom available, a simplified management model was also developed (Davidson et al., 1978a) to calculate the total amount of the various nitrogen forms rather than their concentration distributions. This model is limited to use with homogeneous, well-drained soils, whereas the research model can simulate heterogeneous or layered soil profiles. Neither model considers loss of nitrogen in surface runoff.

The phosphorus cycle is not as complicated as the nitrogen cycle. A major consideration is the equilibrium between (PO_4^{3-}) orthophosphate in solution and that adsorbed by the soil. Because there is no clear distinction between surface adsorption and precipitation in various mineral forms, not all the phosphate removed from solution by the solid phase is readily exchangeable. In runoff, sediment with high phosphate adsorption capacity can remove phosphate from the solution phase. The following nonlinear equation is often used to describe the adsorption isotherm:

$$S = KC^n \tag{1}$$

where S is the concentration of exchangeable phosphate in the soil, C is solution concentration, and K and n are coefficients to be determined. The following papers describe models of phosphorus in the soil system: Enfield and Shaw (1975), Helyar and Munns (1975), and Shah et al. (1975).

1.2.3 Pesticides

Most pesticides are synthetic organic compounds, and as the amounts applied are low, usually between 1 and 2 kg ha^{-1}, their concentrations in runoff and sediment are low. Total seasonal losses are commonly less than 3% of the amount applied (Leonard et al., 1976). Some pesticides are found mainly in the solution phase while others are highly adsorbed by the sediment and organic matter. Techniques such as gas chromatography are needed for the quantitative analysis of pesticide levels, but interference from naturally occurring compounds makes accurate analysis at low levels difficult.

The pesticide content of runoff is influenced by the chemical properties of the pesticide, both the method and timing of its application, the soil properties, crop management methods, and the weather. Summaries of available field data show extreme variability of pesticide mass loads and concentration in runoff, hindering attempts at generalization. It seems as though the pesticide content of runoff can be analyzed only on a site-by-site and event-by-event basis, although this is impractical and some generalization is needed (Haith, 1976).

The chemical or biological degradation of pesticides is the major pathway of their loss, and the predominant mechanisms of loss depend upon chemical and environmental factors. No single equation can describe the degradation of all pesticides in all environment, but a first-order rate equation appears to be an adequate first approximation for many pesticides (Frere, 1975, 1982):

$$A = A_0 \exp\,[-(B_c)t] \tag{2}$$

where A is the amount of pesticide remaining after time t, A_0 is the amount at time zero, and B_c is a breakdown coefficient in reciprocal time units ($B_c = 0.693/t^*$, where t^* is the half-life of the compound). A more detailed approximation writes the breakdown coefficient, B_c, as a function of environmental conditions such as temperature, moisture, and sunlight. Models for movement of chemicals through layered profiles, like those proposed by Davidson et al. (1975), Lindstrom and Boersma (1973), Oddson et al. (1970), and Shah et al. (1975), compute movement from one layer to another. Frere (1975) presented a simple model based on the assumption that the peak chemical concentration in soil lags the water front in proportion to the degree of adsorption of the chemical by the soil. The peak concentration of the chemical is assumed to decrease with the square root of the distance moved by the peak as a consequence of dispersion effects.

The process by which runoff extracts chemicals from the soil is complicated and ill defined, and necessary measurements and experiments have yet to be conducted. One predictive approach suggested by Frere (1975) assumes that leaching and runoff are equally effective in moving chemical. The same algorithm proposed for downward movement of pesticides (Frere, 1975, 1982) with total rainfall and infiltration can then be used to give two concentrations at the soil surface at the start and the end of runoff, assuming a Guassian distribution about the peak. The mean value of the two surface concentrations is assumed to be the mean concentration of chemical in the runoff water. Other models have been proposed by Bruce et al. (1975), Kang et al. (1970), and Donigian and Crawford (1976b).

For any chemical sorbed upon soil, erosion is a major transport process. When chemicals are applied to the soil surface the relative amounts of rill and interrill erosion can significantly affect their loss (Bruce et al., 1975; Frere, 1975; Rugowski and Tamura, 1970). When a chemical is mixed into the soil, the use of a single composite process called "surface erosion" is probably sufficient to describe its transport (Donigian and Crawford, 1976a, 1977).

1.2.4 Organic Matter

The erosion and transport of organic matter is important because organic matter has a large surface area and a concomitant high potential for adsorbing chemicals. Organic matter is usually enriched in eroded sediment, and observed enrichment ratios range from 1.2 to 4.4 (Foster, 1980). Although some organic matter is in the form of removable low-density individual pieces, much of it is adsorbed to clay particles as long chains of carbon molecules. It is mainly contained in aggregates and deposited and transported with the aggregates. While some authors have assumed that organic matter in sediment is enriched in the same ratio that clay is enriched [the algorithm used in the CREAMS (Chemicals, Runoff, and Erosion from Agricultural Management Systems) model], an enrichment ratio of 2 was assumed for organic matter by Khaleel et al. (1979).

1.2.5 Heavy Metals

Before the use of synthetic organic pesticides become widespread, pesticides were inorganic compounds containing copper, lead, and arsenic. Runoff from contaminated lands, municipal sludge, urban areas, and atmospheric fallout from power plants and industries are the source of heavy metal pollution. The heavy metals are highly adsorbed by soil, so sediment is the principal transporting medium. Some modeling of heavy metal transport to streams has been attempted (e.g., Patterson et al., 1976).

1.3 MODEL HISTORY AND CLASSIFICATION

We can distinguish four distinct periods in the development of mathematical models of nonpoint pollution (Novotny, 1986):

1. *The precomputer age or the age of empirical hydrology (1900–1950).* During this period the focus was entirely on water quality with little consideration of such issues as the environmental aspect of water quality. Examples of the models developed during this period are the Green and Ampt (1911) model of infiltration, followed by those of Horton (1939) and Philip (1954);
2. *The transition period of the 1950s.* Hydrologic data collection accelerated during this period but the analysis was slow and costly. Small catchment studies of hydrologic processes began during this period. The U.S. Department of Agriculture Soil Conservation Service (USDA-SCS) developed the SCS Curve Number Model (Mockus, 1964) and the USLE (Wischmeier and Smith, 1965);
3. *The early years of computer use.* During the 1960s computer models began to be developed. The first hydrologic model by Crawford and Linsley (1962), the Stanford Watershed Model (SWM), became the basis for numerous subsequent studies. The most widely used version of the SWM is the Mark IV, which was published in 1966 (Crawford and Linsley, 1966). This has since been incorporated into several large models, in present use, concerned with chemical transport. During the 1970s, as a result of greater access to computers and larger resources being devoted to modeling agencies such as the U.S. Environmental Protection Agency (EPA), USDA, and National Science Foundation (NSF), many models developed. Their field application during the 1970s was of an experimental nature. Comparisons of measured field data and model prediction were occasionally carried out, but the results were not satisfactory. Later field studies were carried out throughout the U.S. to provide a database for the successful application and verification of models (Leonard et al., 1979);
4. *The mid 1970s to date.* With the rapid development of inexpensive microcomputers, the practical user can now use models for routine evaluation. While new model

development has slowed, compared to the 1970s, a number of software companies, consultants, and model creators are now adapting existing models to microcomputers and for wider practical use.

1.3.1 Types of Models

Mathematical models describing the processes of nonpoint pollution can be categorized in numerous ways. One classification distinguishes empirical and physical process models. Regression models and statistical time series models are examples of empirical models. Empirical models are generally simpler and require fewer data than physical process models, but they are difficult to improve, cannot be extended beyond the range of data used in their development, are easily misapplied, and can be misleading about cause and effect. A familiar empirical model is the USLE, a regression equation used to calculate long-term average annual soil loss from small areas. Physical process models can be used to predict responses that are not necessarily observed, assess effects of environmental change, coordinate and structure research, and point the way to develop and improve empirical models. They may or may not require substantial data to develop but extensive research is required to test all models thoroughly.

Most of the recently developed models attempt to piggyback sediment erosion and chemical processes onto detailed hydrologic models. Of the relatively large number of models, only a few become truly operational so that they can be used for management or control purposes.

Nonpoint pollution models fall into two distinct categories: screening models and hydrologic assessment models (Barnwell and Krenkel, 1982; Novotny and Chesters, 1981). Screening models are based on a simple value or function that expresses pollution generation per unit area and unit time for each typical land use. These models are used to identify problem areas in a large basin. While this concept of relating pollution loading to land use is not accurate it is popular because it simply provides quick answers to pollution problems of large areas. More complicated models fail when applied to such a large area because of the enormous amount of information required. This approach is comparable with "overview modeling," whereby unit loadings are combined with information on land use, soil characteristics, and distribution to yield watershed loadings or to identify areas likely to be producing high amounts of nonpoint pollution (Johnson et al., 1978; Sonzogni et al., 1980; Haith and Tubbs, 1981).

Hydrologic assessment models can be divided into:

1. *Physical,* such as a scaled down facsimile of the full-scale prototype (Chow, 1967; Chery, 1963; Clarke, 1973; Novotny, 1986)
2. *Analogue,* such as a complete catchment used by Ishi-hara (1961) and Diskin (1967)
3. *Mathematical or abstract,* in which the behavior of the system is represented by a set of operations together with logical statements expressing relationships between variables and parameters (Clarke, 1973; Freeze, 1975; Novotny, 1986; Rose et al., 1983)

Models of the type described in (1) and (2) are of little significance to nonpoint pollution modeling. The mathematical models to which this review is restricted have been grouped in different ways. Clarke (1973), Freeze (1975), and Frere (1982) have distinguished four major groups:

1. Stochastic — conceptual
2. Stochastic — empirical
3. Deterministic — conceptual
4. Deterministic — empirical

Each of these is then subclassified into linear or nonlinear [(1) and (2) only], probability distributed, lumped, or geographically distributed.

Unfortunately no unanimity exists for the meaning of these terms. One person's theoretical or conceptual model may be someone else's statistical or black-box model.

Lumped parameter models are the most common approach to modeling nonpoint pollution, but some complex models developed recently are based on distributed parameter concepts. Lumped parameter models treat the watershed as one unit. Variations within this unit are lumped together and simplified so that the watershed is represented by the model as a uniform homogeneous system. Distributed parameter models are mostly run for single events or small series of events. They divide watersheds into small, finite elements with uniform characteristics of soil, crop, slope, etc. The mathematical foundation of these models commonly uses basic differential equations governing the flow and mass continuity and motion in one, two, or three dimensions. Lumped models have only one output for a system while distributed parameter models provide an output for each element. With distributed parameter models, area loading maps and graphics can be generated and the effect of changes in watershed management can be easily examined. However, such models require large computer storage and extensive description of parameters for each unit.

Models can be designed to run on a single event or on a continuous basis. Event models require less extensive meteorological data and are easier to implement on small computers. Their limitation is that storm and antecedent moisture condition need to be specified. Event-oriented models are more appropriate for comparative studies of the effects of land management on pollution and water quality. When used to estimate long-term delivery of pollutant to a reservoir they must be integrated with long-term climatic characteristics and, when used in this way, may require some simplification. Alternatively, the outcome of using the full event model in a variety of environmental situations may be investigated and generalized to provide a long-term probability distribution by using long-term climatic data.

Stochastic and probabilistic nonpoint pollution models are less common and mostly confined to research purposes. Bogardi and Duckstein (1978) present an example of a stochastic model of phosphorus loadings from nonpoint sources.

More discussion about different types of models is available in Clarke (1973) and Freeze (1975).

1.3.2 Model Structure

Typical components of a nonpoint pollution model are as follow:

1. A surface runoff component dealing with the formation of runoff and surface flow resulting from rainfall. It may include surface water storage, snow melt, and evapotranspiration.
2. An infiltration and ground-water component describing the movement of water through the unsaturated soil zone into a saturated groundwater zone. The infiltration rate is usually estimated from soil texture, but may be estimated by empirical equations such as the SCS Runoff Curve Number. The most commonly used infiltration models are those of Green and Ampt (1911), Philip (1954), and Holtan (1961).
3. An erosion component that estimates soil loss from pervious areas. The most commonly used model is the USLE (Wischmeier and Smith, 1965). The component for pollutant accumulation and washoff from impervious areas estimates the mass of pollutant from impervious urban and suburban area (Novotny and Chesters, 1981).
4. A soil adsorption/desorption component that determines the distribution of adsorbed and dissolved fractions of pollutants in soils. This component, which is not common to all models, may include volatilization and decay of material such as pesticides and nitrogen.
5. A runoff and sediment routing component. In order to generate runoff flow or pollution histograms the excess rain and eroded soil must be routed overland to the watershed exit. In distributed parameters modeling, routing is accomplished by a kinematic wave model together with the continuity equation. In lumped models some form of transform function is employed (Nash, 1957). It is expected that their predictive use in a stochastic or probabilistic long-term context will increase.

1.4 CHARACTERISTICS IMPORTANT IN CHOOSING MODELS

1.4.1 How Stochastic Issues Are Dealt with

There are a variety of ways in which stochastic descriptions may be used in models. First, there is uncertainty in the driving variables used in models, such as climatic factors. Such input variables (e.g., rainfall amount or rate) may be expressed in terms of a location-specific probability distribution. While such an input requires stochastic representation to describe expected long-term behavior, processes at the soil surface still may be described in a deterministic manner. Use of stochastic description in this way simply recognizes that if a model is to be used predictively to assess outcomes in the context of real-world uncertainty in driving variables, then all that can be done is to represent the probability structure of that real world accurately.

Such use of stochastic inputs can be avoided only if a model is designed to accept particular historic data and to analyze the observed outcomes associated with that data. This is the common mode of operation at the stage of model development and testing. When, having been evaluated in such a way, the model is accepted as of useful validity, then a common model objective is to expand the time frame of data collection by interacting the model with longer term climatic data than the limited suite of climatic experience in which the model has been tested.

In modeling, decisions must be made on the level of detail to be used in representing the physical, chemical, or biological systems of relevance. The method of system description may be dominated by current understanding of the system processes, in which case the type of model would usually be categorized as deterministic. Alternatively it may be decided not to attempt such a process description, or only to represent system outcomes obtained by an analysis of such outcomes using mathematical/statistical analysis that facilitates a description of such process outcomes, perhaps in terms of some distribution. Such a model provides a mathematical description of the outcome of system behavior that may or may not be at all dependent on any representation of system structure or processes. The particulars of such model are therefore captive to the set of data describing such behavior, although the model structure may well have generality of application. However, if this data set is extensive, this may not be a major limitation unless one wishes to transfer use of the model to another context.

The more a model is based on the particular outcomes of system behavior, rather than more basic descriptions of system processes, the more likely it is that the model is location specific in its prediction, and so the greater the caution that should be used in transferring the model for use in other contexts. This is not to say that such a modeling methodology cannot be used elsewhere, but that its use will require the existence or development of a similarly extensive database of system outcomes.

If such an extensive database already exists or is rather readily obtained, then summarizing that database is more appropriate and may be the most efficient to adopt. However, the development of such a database for a range of chemicals (or even a single chemical) even in a single catchment is commonly a difficult matter, and long-term records of chemical output do not usually exist. In addition, changes in land use or management may make the direct use of system behavior questionable.

For such reasons there is a commonly implicit assumption in the literature on modeling chemical delivery from catchments to streams that it is desirable to adopt modeling structures that bear some resemblance to current knowledge of processes, even if at a somewhat generalized or simplified level. Such simplification is favored usually because of a recognition of the great complexity in what is being represented, and that it is not feasible to represent such complexity in any completeness.

The nature of representation of physical and chemical processes and how they are represented also affect model transferability. If a process of significance is going on in a watershed but is not represented in the model, then the model is not suitable for representing that process or other processes affected by it.

As experience and confidence with such deterministic models increases, then it is expected that their predictive use in a stochastic or probabilistic, long-term context will increase.

1.4.2 Errors

A knowledge of likely errors and the sources of uncertainty is important in model use. The major sources of error are model error, input error, output error, and errors caused by spatial or time variability.

Common sources of model error include (1) modeling a distributed system in a lumped form, (2) omitting processes that are important but thought not to be of interest, and (3) representing nonlinear processes as linear. Model error is always present because of the abstract nature of modeling.

Input error is a second major source of error, especially in hydrologic modeling, because one of the main inputs, precipitation, can be quite variable over an area, both in total amount and in its rate history. If erroneous data are used to calibrate the model, the estimated parameter will also contain error. Input error also makes estimation of model error more difficult because the difference between model prediction and observed response of watershed can partly be due to input error. Output error is less serious. Its major source in small watershed chemical transport comes from inadequate frequency in sampling or by unrepresentative sampling.

1.4.3 Spatial Variability

Physical, chemical, and biological characteristics are nonuniform across a watershed. When different soils and land uses are considered as one unit, then it is not clear what determines the integrated value observed at the outlet. One approach adopted to overcome spatial variability is to use cumulative frequency distribution of a characteristic or process across the watershed. The Stanford Model uses linear approximations to the cumulative frequency distribution of infiltration, while the Agricultural Runoff Management (ARM) model (discussed in Section 2.5.3) arbitrarily divides watersheds into five equal parts based on infiltration capacity. The USDAHL model (discussed in Section 2.2.5) divides the watershed into three to four zones of approximately uniform hydrologic response. The model developed by Colorado State University (Section 2.3.12) uses a grid system to divide the watershed into subunits. ANSWERS (Areal, Nonpoint Source Watershed Environment Response Simulation) (discussed in Section 2.3.11) conceptually divides watersheds into squares of uniform characteristics.

1.4.4 Time Variability

Variability in time can be another source of error because large time intervals between sampling nonlinear processes can introduce very significant error, especially in models based on measurement data.

Continuous or frequent collection and analysis of samples is difficult or expensive. Frequency of sampling is also involved in the ability to detect widely different time scales such as diurnal, storm, or seasonal variability of a system.

1.5 REFERENCES

Barnwell, T. O. and P. A. Krenkel. 1982. The use of water quality models in management decision making. *Water Sci. Technol.* 14:1095–1107.

Bogardi, I. and L. Duckstein. 1978. Input for stochastic control model of P loading. *Ecol. Model.* 4:173–195.

Bruce, R. R., L. A. Harper, R. A. Leonard, W. M. Snyder, and A. W. Thomas. 1975. A model for runoff of pesticides from small upland watersheds. *J. Environ. Qual.* 4(4):541–548.

Chery, D. L. 1963. Runoff studies with a physical model of a watershed. *Trans. Am. Geophys. Union* 44:869–870.

Chow, V. T. 1967. Laboratory studies of watershed hydrology. *Proc. Int. Hydrol. Symp.* 1:194–202.

Clarke, R. T. 1973. A review of some mathematical models used in hydrology with observation on their calibration and use. *J. Hydrol.* 19:1–20.

Crawford, N. H. and R. K. Linsley. 1962. The synthesis of continuous streamflow hydrographs on a digital computer. Technical Report No. 12. Department of Civil Engineering, Stanford University, Palo Alto, CA.

Crawford, N. H. and R. K. Linsley. 1966. Digital simulation in hydrology. Stanford watershed model IV. Department of Civil Engineering, Stanford University, Palo Alto, CA.

Davidson, J. M., G. H. Brusewitz, D. R. Baker, and A. L. Wood. 1975. Use of soil parameters for describing pesticide movement through soils. EPA-660/2-75-009. EPA Technological Service, U.S. Environmental Protection Agency, Athens, GA.

Davidson, J. M., D. A. Graetz, P. S. C. Rao, and H. M. Selin. 1978a. Simulation of nitrogen movement, transformation and uptake in plant root zone. EPA-600/3-78-029. U.S. Environmental Protection Agency, Athens, GA.

Davidson, J. M., L. T. Ou, and P. S. C. Rao. 1978b. Adsorption, movement and biological degradation of high concentrations of selected pesticides in soils. In: Proc. 4th Annual Research Symposium: Land Disposal of Hazardous Wastes. EPA-600/9-78-016. U.S. Environmental Protection Agency, Athens, GA, pp. 233–244.

Diskin, M. H. 1967. A dispersion analogue model for watershed systems. *Proc. Int. Hydrol. Symp.* 1:38–45.

Donigian, A. S., Jr. and N. H. Crawford. 1976a. Modelling nonpoint pollution from the land surface. EPA-600/3-76-083. EPA Technological Service, U.S. Environmental Protection Agency, Athens, GA.

Donigian, A. S., Jr. and N. H. Crawford. 1976b. Modelling pesticides and nutrients on agricultural lands. Report No. EPA-600/2-76-043. U.S. Environmental Protection Agency, Athens, GA.

Donigian, A. S., Jr. and N. H. Crawford. 1977. Simulation of nutrient loading in surface runoff with the NPS model. EPA-660/3-77-065. U.S. Environmental Protection Agency, Athens, GA.

Enfield, C. G. and D. C. Shaw. 1975. Comparison of two predictive nonequilibrium one-dimensional models for phosphorous sorption and movement through homogenous soils. *J. Environ. Qual.* 4(2):198–202.

Foster, G. R. 1980. Soil erosion modelling: special considerations for nutrient pollution evaluation of field sized area. In: Environmental Impact of Nonpoint Source Pollution. M. R. Overcash and J. M. Davidson (Eds.). Ann Arbor Science, Ann Arbor, MI, pp. 213–243.

Freeze, R. A. 1975. A stochastic-conceptual analysis of one-dimensional groundwater flow in nonuniform homogeneous media. *Water Resour. Res.* 11:725–741.

Frere, M. H. 1975. Integrating chemical factors with water and sediment transport from a watershed. *J. Environ. Qual.* 4(1):12–17.

Frere, M. H. 1982. Modelling the quality of water from agricultural land. In: Hydrologic Modelling of Small Watersheds. ASAE Monograph No. 5. Haan et al. (Eds.). American Society of Agricultural Engineers, St. Joseph, MI, pp. 383–401.

Green, W. H. and G. A. Ampt. 1911. Studies in soil physics. 1. The flow of air and water through soils. *J. Agric. Sci.* 4:1–24.

Haith, D. A. 1976. Land use and water quality in New York River. *J. Environ. Eng. Div. Am. Soc. Civ. Eng.* 102:1–15.

Haith, D. A. and L. J. Tubbs. 1979. Modelling nitrogen export in rainfall and snowmelt runoff. In: Best Management Practices for Agriculture and Silviculture. R. C. Loehr, D. A. Haith, M. F. Walter, and C. S. Martin (Eds.). Ann Arbor Science, Ann Arbor, MI, pp. 665–685.

Haith, D. A. and L. J. Tubbs. 1981. Watershed loading functions for nonpoint sources. *J. Env. Eng. Div. ASCE* 107:121–137.

Helyar, K. R. and D. N. Munns. 1975. Phosphorus flux in the soil plant system: a computer simulation. Hilgardia 43(4):103–130.

Holtan, N. H. 1961. A concept for nitrification estimates in watershed engineering. Report No. ARS-41-51. Agricultural Research Service, U.S. Department of Agriculture, Washington, D.C.

Horton, S. E. 1939. Estimating soil permeability rates. *Proc. Soil Sci. Soc. Am.* 5:399–417.

Ishi-hara, T. 1961. Runoff analysis by analogue computer. In: General Assembly of the International Association for Hydraulic Research, 9th, Dubrovnik, pp. 749–759.

Johnson, M. G. et al. 1978. Management information base and overview modelling. PLUARG report. International Join Commission U.S.–Canada. Windsor, Ontario, Canada.

Kang, F. F., L. T. Fan, E. S. Lee, and L. E. Erickson. 1970. Modelling feedlot runoff pollution. 1. Analogue simulation. *Trans. ASAE* 13:859–863.

Khaleel, R., G. R. Foster, K. R. Reddy, M. R. Overcash, and P. W. Westerman. 1979. A nonpoint source model for land areas receiving animal wastes. IV. Model inputs and verification for sediment and manure transport. *Trans. ASAE* 22:1362–1368.

Leonard, R. A., G. W. Bailey, and R. R. Swank, Jr. 1976. Transport, detoxification, fate and effects of pesticides in soil and water environments. In: Land Application of Waste Materials. Soil Conservation Society of America, Ankeny, IA, pp. 48–78.

Leonard, R. A., G. W. Longdale, and W. G. Flemming. 1979. Herbicide runoff from upland piedmont watersheds — data and implication for modeling pesticide transport. *J. Environ. Qual.* 8:223–229.

Lindstrom, F. T. and L. Boersma. 1973. A theory on the mass transport of previously distributed chemicals in water saturated, sorbing porous media. 111. Exact solution for first order kinetic sorption. *Soil Sci.* 115(1):5–10.

Mockus, V. 1964. *SCS National Engineering Handbook.* Section 4, Hydrology. U.S. Soil Conservation Service, 548 pp.

Moldenhauer, W. C. and C. A. Onstad. 1975. Achieving specific soil erosion control. *J. Soil Water Conserv.* 30(4):166–168.

Nash, J. E. 1957. The form of the instantaneous unit hydrograph. *Bull. Int. Assoc. Sci. Hydrol.* 111:114–121.

Novotny, V. 1986. A review of hydrologic and water quality models used for simulating agricultural pollution. In: Agricultural Nonpoint Source Pollution, Model Selection and Application. *Dev. Environ. Model.* 10:9–35.

Novotny, V. and G. Chesters. 1981. Handbook of Nonpoint Pollution: Sources and Management. Van Nostrand Reinhold, New York.

O'Conner, D. J., R. V. Thonnan, and D. N. Ditoro. 1973. Dynamic water quality forecasting and management. EPA-660/3-73-009. EPA Technological Service, U.S. Environmental Protection Agency, Athens, GA.

Oddson, J. K., J. Letey, and L. V. Weeks. 1970. Predicted distribution of organic chemicals in solution and adsorbed as a function of position and time for various chemical and soil properties. *Soil Sci. Soc. Am. Proc.* 34(3):412–417.

Patterson, M. R., C. L. Beguvich, and D. R. Jackson. 1976. Environmental transport modelling of pollutants in water and soil. In: Symposium on Nonbiological Transport and Transformation of Pollutant on Land and Water. National Bureau of Standards, Gaithersburg, MD.

Philip, J. R. 1954. An infiltration equation with physical significance. *Soil Sci.* 77:153–157.

Rose, C. W., F. W. Chichester, J. R. Williams, and J. T. Ritchie. 1982a. A contribution to simplified models of field solute transport. *J. Environ. Qual.* 11:146–150.

Rose, C. W., F. W. Chichester, J. R. Williams, and J. T. Ritchie. 1982b. Application of an approximate analytic method of computing solute profiles with dispersion in soils. *J. Environ. Qual.* 11:151–155.

Rose, C. W. and R. C. Dalal. 1988. Erosion and runoff of nitrogen. In: Advances in Nitrogen Cycling in Agricultural Ecosystems. J. R. Wilson (Ed.). Commonwealth Agricultural Bureau International, Wallingford, England.

Rose, C. W., J.-Y. Parlange, G. C. Sander, S. Y. Campbell, and D. A. Barry. 1983. A kinematic flow approximation to runoff on a plane: an approximate analytic solution. *J. Hydrol.* 62:363–369.

Rugowski, A. S. and T. Tamura. 1970. Erosional behaviour of cesium-137. *Health Phys.* 18:467–477.

Shah, D. B., G. A. Coulman, L. T. Novak, and B. G. Ellis. 1975. A mathematical model for phosphorus movement in soils. *J. Environ. Qual.* 4(1):87–92.

Smith, C. N. 1980. Sediment-pollutant relationships in runoff from selected agricultural, suburban and urban watersheds. EPA-600/3-80-022. U.S. Environmental Protection Agency, Athens, GA.

Smith, S. J., L. B. Young and G. E. Miller. 1976. Field evluation of soil nitrogen mineralization potential. *Soil Sci. Soc. Am. Proc.* 40.

Sonzogni, W. C. et al. 1980. WATERSHED: A management technique for choosing among point and nonpoint control strategies. In: Proc. of the Seminar on Water Quality Management. U.S. Environmental Protection Agency, Chicago, IL.

Wischmeier, W. H. and D. D. Smith. 1965. Predicting rainfall-erosion losses from cropland east of Rocky Mountains — guide for selection of practices for soil and water conservation. Agriculture Handbook No. 282. U.S. Department of Agriculture, Washington, D.C.

Young, R. A. and C. A. Onstad. 1976. Predicting particle size composition of eroded soil. *Trans. ASAE* 19(6):1071–1075.

CHAPTER 2

Sorbed Chemical Transport Modeling

H. Ghadiri and C. W. Rose

2.1 INTRODUCTION

Classification of nonpoint source models into distinct categories is not possible. However, to reflect common model structures, this chapter groups models into those concerned with hydrology, with sediment transport, and with chemical transport. Generally a chemical transport model will contain hydrologic and sediment transport model components, and a sediment transport model with a hydrologic submodel. Therefore in this chapter (with some exceptions), this review proceeds from simple or single-process models to more complicated multifunctional models.

An attempt has been made to cover a wide selection of the agricultural nonpoint source pollution models developed in recent years. Urban and receiving water-quality models have been treated only briefly. Some models are not as fully described or critically reviewed as others due to the unavailability of the original publications. For such models we therefore present only a general overview of their structure and function, and references to further sources of information about them.

Some comparative studies of the models on sorbed chemical transport are covered in Section 2.6. All the models reviewed in this chapter and the references consulted for their review have been summarized in Table 2.1.

2.2 HYDROLOGIC MODELING

2.2.1 Introduction to Hydrologic Modeling

Overland flow is the mechanism of chemical transport considered in this review. The two requirements for the generation of overland flow are that the rate of rainfall

ISBN 0-87371-747-3

Table 2.1 Reviewed Models and Their References

Number	Model	Ref.
1	ACTMO	Frere et al. (1974, 1975)
2	AGNPS	Young et al. (1987)
3	AGRUN	Donigian et al. (1977)
4	ANSWERS	Beasley and Huggins (1980); Beasley et al. (1980)
5	ARM	Donigian and Crawford (1976a,b); Donigian et al. (1977)
6	CNS	Haith and Loehr (1979)
7	CPM	Haith and Loehr (1979)
8	CREAMS	Knisel (1980)
9	CREAMS-WT	Heatwole et al. (1987, 1988)
10	Curtis	Curtis (1976)
11	Erosion model of Yang, Moore and Burch	Yang (1973); Moore and Burch (1986)
12	EPARRB	True (1976)
13	EPIC	Williams (1983)
14	GAMES	Dickinson et al. (1984); Cook et al. (1986)
15	GAMESP	Rousseau et al. (1985)
16	GLEAMS	Leonard et al. (1987)
17	GUEST	Misra and Rose (1989)
18	HILLS	Smith and Hebbert (1983)
19	HSPF	Donigian et al. (1983); Donigian (1986)
20	Hydrologic Del. contributing areas	Gburek (1983)
21	IUSG	Kumar and Rastogi (1987)
22	Loading functions	Davidson et al. (1978a,b); Leonard et al. (1979); Rao and Davidson (1979); Baker (1980); Haith (1980); Logan (1980); Dean (1983)
23	KINCON	Silburn and Loch (1990)
24	KYERMO	Hirschi and Barfield (1988a,b)
25	Model for land receiving animal waste	Khaleel et al. (1979a,b); Reddy et al. (1979a,b)
26	MRI	McElroy et al. (1976)
27	NONPT	Holtan and Lopez (1971); Holtan (1979)
28	NTRM	Shaffer et al. (1983)
29	NPS	Donigian and Crawford (1976a, 1977)
30	PERFECT	Littleboy et al. (1989)
31	PLIERS	Kenimer et al. (1989)
32	PWM	Akan (1987)
33	PRM	Haith (1980, 1986)
34	PRMS	Leavsley et al. (1983)
35	RORB	Laurensen and Mein (1985)
36	SCS-CNE	Mockus (1964)
37	SEDIMOT	Wilson et al. (1986)
38	Soil erosion model of Foster	Foster (1982)
39	Soil erosion model of Morgan et al.	Morgan et al. (1984)
40	SOILOSS	Rosewell and Edwards (1988)
41	SPAW	Saxton et al. (1984)
42	SPNM	Williams (1980)
43	SPUR	Lane (1982); Wight (1983)
44	SWM	Crawford and Linsley (1966)
45	STORM	U.S. Army Corps of Engineers (1977)
46	SWAM	DeCoursey (1982)

Table 2.1 Reviewed Models and Their References (continued)

Number	Model	Ref.
47	SWMM	Metcalf and Eddy, Inc. (1971); Huber et al. (1981)
48	SWRRB	Williams and Nicks (1985); Williams et al. (1985)
49	The Colorado State University models	Li (1974); Simons et al. (1975, 1977); Simons and Li (1980)
50	TOPOG	Vertessy et al. (1990)
51	USDAHL	Holtan and Lopez (1971); Holtan et al. (1975, 1977)
52	USLE	Wischmeier and Smith (1965, 1978)
53	UTM	Huff et al. (1977)
54	WASCH	Bruce et al. (1975)
55	Water erosion model of Rose and Hairsine	Rose and Hairsine (1988)
56	WEPP	Foster (1987); Lane et al. (1988)
57	WEST	Leytham and Johanson (1979)

exceeds the rate of infiltration, and the surface detention storage is satisfied. The amount of surface detention depends upon the micromorphology of the soil surface, which can be modified by erosion.

Rainfall rates can exceed infiltration rate in two quite different contexts. If the rainfall rate is sufficiently high, ponding can be achieved with the soil surface profile becoming saturated, beneath which there may exist a zone transmitting water to a wetting front. Generation of overland flow in this way is associated with the name of Horton, and the time for ponding to occur is related to the sorptivity and rate of rainfall (Perroux et al., 1981). The generation of runoff in such situations is not directly related to landscape topography, but depends chiefly on the ratio of rainfall rate and sorptivity, being more common if this ratio is high. The second and contrasting context for the generation of overland flow is in the zone of three-dimensional landscapes that have become saturated so that the subsurface downslope water flux exceeds the ability of the soil profile to transmit that flux. A waterlogged zone forms whose geometry and general location is determined by the three dimensionality of the landscape (O'Loughlin, 1981, 1986). The extent of the saturated zones also depend on soil profile transmissivity. Overland flow is then generated by rain falling on this saturated zone. This context of overland flow generation is common in more temperate regions with rates of rainfall generally lower than those typical of the tropics and semitropics.

The formation of zones of saturation are implicit in the numerical analysis of interdependent surface and subsurface hydrologic processes. At least if steady-state assumptions are made, saturation zones can be located much more simply using the approach pioneered by O'Loughlin (1981). A major advantage of this model, if its assumptions are satisfied, is that a complete solution of subsurface water movement in all three-dimensional space of the catchment is not necessary. In this approach the subsurface water flux of infiltrated rainfall is accumulated in flow wedges whose lateral boundaries are orthogonal to contours. Saturation occurs when this accumulated lateral flux in the soil profile above an assumed impermeable lower boundary exceeds the ability of the profile to transmit it. This type of analysis becomes practically feasible due to digital terrain-mapping capability. The methodology has been applied, with considerable success, to simulate changes in waterlogged areas following forest clearing (e.g., Moore et al., 1986; O'Loughlin, 1986).

In principle, all real-world generation of overland flow would seem to lie between the two limits of Hortonian flow (Horton, 1939) and runoff from topographically developed saturation zones. Within a catchment there can be spatial variation in time to ponding due either to variation in infiltration rate characteristics or to differences in the capacity of the profile to store water. Van de Griend and Engman (1985) have outlined the difficulties in using this concept in practical hydrology. Boughton (1987) has described a practical method by which the proportion of a catchment contributing to overland flow can be determined by analysis of rainfall and runoff records. It assumes spatial variability in storage capacity, but would also apply if spatial variability was in infiltration characteristics. The analysis of surface saturation zones of topographic analysis and flow routing where appropriate (Moore et al., 1986) would be expected to provide deeper physical insight into variable source areas of runoff.

Whether overland flow is extensive (or Hortonian) or restricted to time-varying source areas of saturated soil, the question remains as to the dynamic relationship between runoff and excess rainfall rate. For a plane land surface this problem has been solved analytically for any time-variant excess rainfall rate (Parlange et al., 1981). A simpler approximation (Rose et al., 1983b) is of the form:

$$R_e = Q + k_1\, dQ/dt \quad \left(\mathrm{m\ s^{-1}}\right) \tag{1}$$

where R_e is excess rainfall rate (assumed uniform over the plane), Q is runoff rate per unit area of plane, and k_1 is a known constant involving the length, slope, and roughness of the plane. (Surface roughness can be due to soil structure, microrelief, plant residue, etc.) By definition:

$$R_e = R_r - I \quad \left(\mathrm{m\ s^{-1}}\right) \tag{2}$$

where I is the infiltration rate and R_r is the rainfall rate. Equations (1) and (2) can be used to determine I as a function of time under rainfall falling on a plane whose slope and length (or strictly the length of ponding contributing to runoff) are known, and where $Q(t)$ is measured. $R(t)$ can then be calculated from Equation (1) if surface ponding can be neglected and, with rainfall rate also measured, $I(t)$ follows from Equation (2). This methodology has the advantages over other methods of providing $I(t)$ under natural rainfall and of providing an appropriately spatially averaged value for the area contributing to the measured runoff. Campbell et al. (1984) showed that Equation (1) also applies for the uniformly diverging or converging surfaces of a cone or inverted cone provided the appropriate analytical expressions are used for k_1.

The approximate analytical solution of Equation (1) is sufficiently simple that it can be readily incorporated into rainfall-runoff models without requiring extensive computer time and Moore (1985) suggests it is therefore attractive in catchment modeling studies. Moore (1985) and Moore and Kinnell (1987) have improved the accuracy of Equation (1) by generalizing a water profile shape factor, assumed constant in Rose et al. (1983a), with essentially no increase in computational complexity. Shape factors for converging and diverging planar surfaces have also been derived (Moore, 1985). Similar correcting shape factors for conical or inverted conical surfaces have yet to be derived.

It is interesting that an equation of form similar to Equation (1) has been developed and used in determining the flood response of urbanized catchments by Bufill and Boyd (1987). Our foregoing discussion has focused on overland sheet flow on land surfaces. Moore and Burch (1986) have shown how the modification of sheet flow by its channeling into rills of known geometry and frequency can be dealt with making the same assumptions of kinematic flow and Manning's equation as are commonly made to describe sheet flow.

In temperate climates the maximum rates of runoff may occur with snowmelt, not rainfall. Processes affecting the rate of snowmelt are not dealt with in this review. Typically upland runoff concentrates into flows, whether or not this is guided by man-made structures such as graded terraces. Water needs to be routed through networks to the more permanent channel or stream systems. Again the choice for modeling these networks is between analytical or numerical (Brakensiek et al., 1966) methods.

While computers are used to implement the models outlined above, not all have been developed in a form where they can be readily used by others. Some examples of those available for general use will now be given.

2.2.2 SCS-CNE

The U.S. Soil Conservation Service Curve Number Equation (US SCS-CNE) method of estimating direct runoff from storm rainfall was developed in the mid-1950s (Mockus, 1964) and is designed for use with rainfall and watershed data that are readily available or easily obtainable for ungauged watersheds (Rallison, 1980). It still forms the basic hydrologic component of many U.S. Department of Agriculture (USDA) models such as CREAMS (Chemical, Runoff, and Erosion from Agricultural Management Systems) and SPUR (Simulation of Production and Utilization of Rangelands) described in Sections 2.5.6 and 2.3.18, respectively.

Its principal application is in estimating runoff quantities in flood hydrographs or in relation to flood peak rates on small to medium-sized catchments. Runoff consists of one or more of the four types: channel runoff, surface runoff, subsurface flow, and base flow. Channel runoff occurs when rain falls on a flowing stream or on an impervious surface and is generally of negligible quantity. Surface runoff occurs when the rainfall rate is greater than the infiltration rate. It appears on the hydrograph after initial demands of interception, infiltration, and surface storage have been satisfied. Subsurface flow occurs when infiltrated runoff meets a low transmission layer and starts traveling downhill above the layer. It emerges as a seep or spring and appears on the hydrograph during or soon after the storm. Base flow occurs when there is a steady flow from natural storage such as lakes, swamps, or replenishable aquifers. It appears very late after a storm and has little effect on a storm hydrograph, although the base flow from previous storms may increase the flow rate.

The SCS method deals with the base flow separately, but combines the other three types in "direct runoff." It estimates direct runoff, but the proportion of surface runoff to subsurface flow can be obtained by means of curve number (ignoring channel flow).

The most generally available rainfall data are the amounts of rainfall measured at nonrecording rain gauges. The SCS rainfall-runoff relation was developed for use with these data (Mockus, 1964). The data are rainfall totals arising from one or more storms occurring in a day and nothing is known about the time distribution of the rainfall. The timing and intensity of rain cannot be treated as separate variables and are ignored. The rainfall-runoff equation is

$$R_a = \frac{\left(Q_m - I_a\right)^2}{\left(Q_m - I_a\right) + S_m} \tag{3}$$

where R_a is the annual rainfall, Q_m is the potential maximum runoff, I_a is the initial abstraction, and S_m is the potential maximum retention (S_m includes I_a). The initial abstraction consists mainly of interception, infiltration, and surface storage, all of which occur before runoff begins. By assuming that the initial abstraction is equal to 20% of the potential maximum retention ($I_a = 0.2S_m$), the above equation can be simplified to

$$R_a \frac{\left(Q_m - 0.2S_m\right)^2}{Q_m + 0.8S_m} \tag{4}$$

which is the rainfall-runoff relation used in the SCS method of estimating direct runoff from storm rainfall. The retention parameter S_m is related to a curve number, CN, by the relationship

$$\mathrm{CN} = \frac{1000}{S_m + 10} \tag{5}$$

Graphs and tables have been prepared for the quick solution of the SCS Curve Number Equation (Mockus, 1964, 1972). Graphical representation of Equation (4) is given in Figure 2.1.

A detailed critical review of the SCS Curve Number method has recently been carried out by Boughton (1989), who argues that the main source of error in estimating runoff by this method is the selection of the curve number. A change of

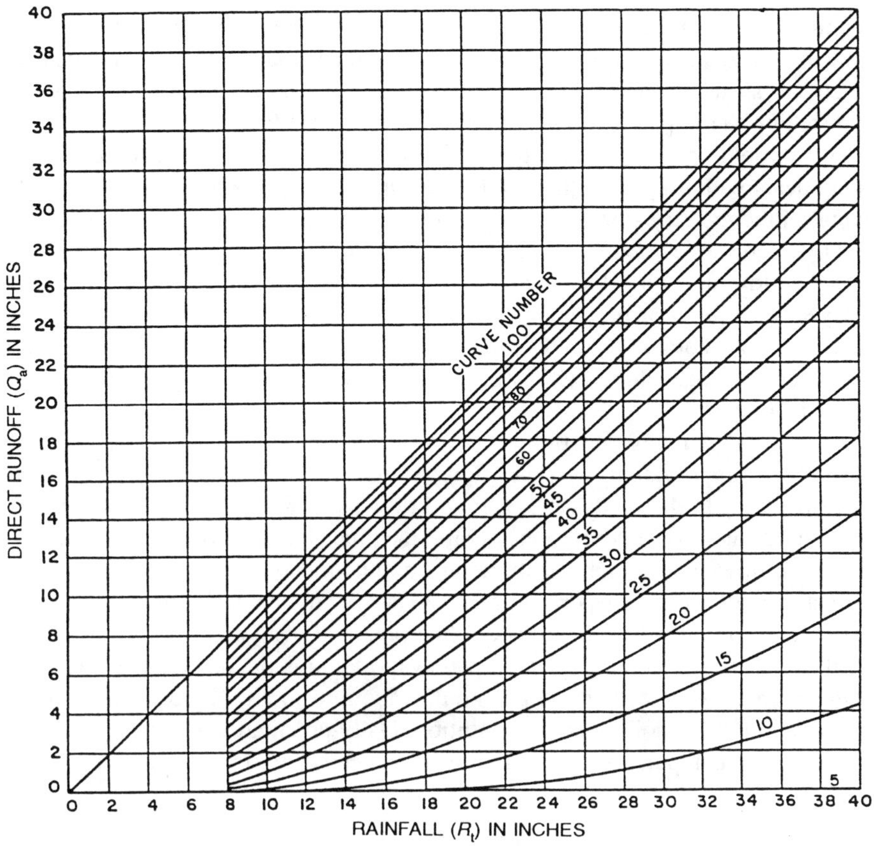

FIGURE 2.1. Solution of rainfall runoff equation $R_a = (Q_m - 0.2S_m)/(Q_m + 0.8S_m)$. (From Mockus, 1956.)

about 15% in the curve number doubles or halves the total estimated runoff. Even after calibrating the curve number with several years of data, error can still occur. The lack of any physical basis for the rainfall-runoff equation is an inherent limitation that limits its potential for further development.

2.2.3 SWM

One of the most widely used hydrologic models is the Stanford Watershed Model (SWM) developed in 1962 and revised in 1966 (Crawford and Linsley, 1966). It requires a large-scale, digital computer system to analyze hydrologic regimes quantitatively, using mathematical relationships between elements of the hydrological cycle.

The flow calculation is divided into land surface and channel flow phases. The land surface flow phase considers interception, impervious areas, infiltration, overland and groundwater flow, and evapotranspiration. Interception is governed by the watershed cover and the current volume in the interception storage. All incoming moisture enters the interception store until a preassigned volume is fitted. The surfaces of lakes and streams, and impervious lands adjacent to streams, contribute directly to surface runoff and are represented in the model by a preset parameter representing the impervious fraction of the total watershed area.

The complex process of infiltration is the key process at the land surface. The model deals with it by calculating the direct infiltration into the soil profile and the increase in temporary storage that delay infiltration. Moisture that is not infiltrating directly increases surface detention storage, which in turn either contributes to overland flow and interflow, or enters upper zone storage (delayed infiltration). Delayed infiltration or percolation occurs from the upper zone to the groundwater and lower zone storages when the upper zone storage ratio (UZS/UZSN) exceeds the lower zone storage ratio (LZS/LZSN) (UZSN and LZSN being terminal storage capacities).

The Chezy-Manning equation is used to relate surface detention storage at equilibrium, the supply rate to overland flow, Manning's n, and the length and slope of the flow plane. In the model, the overland flow hydrographs are added to interflow and groundwater flow hydrographs to form the total channel inflow hydrograph. This channel inflow hydrograph is, therefore, a function of land surface and rainfall characteristics, and is independent of channel system characteristics.

In the evapotranspiration submodel, the potential evapotranspiration is assumed equal to lake evaporation, for which pan evaporation is the practical surrogate. The relationship between actual and potential evapotranspiration in large areas is taken as a function of the soil moisture condition. Daily potential evapotranspiration data (E_p) are used as input, and the actual evapotranspiration, E_a, is calculated:

$$E_a = E_p - \frac{E_p^2}{2r} \tag{6}$$

where the variable r is a factor of the lower zone storage ratio (LZS/LZSN).

The land surface components of the model produce continuous overland flow, interflow, and groundwater flows that enter the stream channel system. Some equations are then used to simulate the time delay and attenuation of channel inflow as it moves through the channel system.

A modified version of Clark's (1945) empirical instantaneous unit hydrograph is used to estimate the outflow hydrograph, neglecting storage attenuation. The flow time from any point in the channel system to the outlet necessary for the construction of a time-delay curve or histogram is calculated using Manning's equation.

The volume of channel inflow in any time interval is multiplied by successive elements of the time-delay histogram to give a watershed outflow hydrograph, and this is routed through a storage system to simulate attenuation in the channel system.

The model has a snowmelt subroutine that assumes that melt can be estimated from the temperature differential between the atmosphere and the snow surface alone. Because the radiant energy exchange is more important to snowmelt than the temperature difference, the output of this subroutine is not very accurate. Another snowmelt subroutine capable of using observed data for all the melt components, i.e., radiation, wind velocity, temperature, and dewpoint, has also been developed by

Anderson and Crawford (1964), but the input requirements are so stringent that its use is limited to well-instrumented watersheds.

2.2.4 RORB

Program RORB illustrates a method of modeling the relationship between runoff and rainfall in rural and urbanized catchments (Laurensen and Mein, 1985). In this program hydrographs are computed by deducting infiltration and storage "losses" from rainfall. The rainfall excess is then routed through a model of the catchment storages. These include both overland flow and channel storage. Storage (S_t) is assumed proportional to the discharge (q) from a plane or channel reach raised to a power (p) so that:

$$S_t = bq^p \tag{7}$$

where both b and p are determined by fitting experimental data, usually on the complete catchment (Laurensen et al., 1985).

The assumption of kinematic flow (i.e., of a unique relationship between flow rate and flow depth) has been widely explored (Lighthill and Whitham, 1955; Woolhizer and Liggett, 1967). Such analysis has usually involved numerical solution of the characteristic equations obtained, although an exact analytical solution is available for a plane land surface or channel (Parlange et al., 1981), and simple approximate analytical solutions for a variety of surface forms (Rose et al., 1983b).

What has not been as widely achieved is the linkage of theoretically rich approaches of the types outlined in the preceding paragraph with the experimentally rich applications of practical rainfall-runoff models such as RORB. Such linkage illustrates, for example, that the empirical relationships between water storage on watershed elements and discharge can be expressed theoretically. For example, S_t, the storage per unit area of plane, is given by:

$$S_t = C_f\left(a/k_1\right)^{1/m} \tag{8}$$

where C_f is a water surface profile shape factor (Moore, 1985), q is the water flux per unit width at exit from the plane, and k_1 and m are defined by the kinematic rating equation

$$q = k_1 D^m \tag{9}$$

where D is depth of flow, and $k_1 = S^{1/2}/8$ where S is plane slope and n Manning's roughness coefficient. Thus, in this simple situation of a plane, the empirical coefficients b and p in Equation (7) can be predicted from Equations (8) and (9).

As an illustration of the utility of making the connection between theory and practice, in fitting hydrograph data for two catchments in Australia, Laurenson et al. (1985) drew attention to the common problem of interaction between the two coefficients b and p in Equation (7), and the consequent difficulty in choosing unique values for either. Comparing Equations (7) and (8) it follows that $p = 1/m$, with $m \approx 5/3$ for turbulent flow and 3 for laminar flow. Laurenson et al. (1985) explored the range 1.0 to 0.5 for p, corresponding to the range 1 to 2 for m. If flow was fully turbulent then $m = 1.7$ could be assumed. Similar comparisons could be made between the empirical factor b and its equivalent, for a single plane, of $C_f / k_1^{1/m}$. The slope and roughness dependence of k_1 would clearly be a cause for variation in the parameter b. Thus the value of b obtained by fitting the watershed hydrograph would depend in part on an integration of the effect of the variety of slopes and roughnesses in that fraction of the catchment yielding overland flow.

2.2.5 USDAHL

USDAHL is a watershed hydrology model developed in 1965 (Holtan, 1965) and revised in 1971 and in 1975 (Holtan, 1971; Holtan et al., 1975, 1977). The revised versions, called USDAHL-74 and USDAHL-74/UMO, have the same hydrologic concepts but their computational efficiency is improved. In this model the watershed is divided into three hydrologic response zones, each described by a unique set of parameters affecting hydrology (Figure 2.2). Zoning can be determined by any physiographic characteristics of interest within the watershed, most often by soil type, topography, or land use. The present model structure uses up to 12 size zones and can be modified to use more if desired. Infiltration is computed by Holtan's (1965) infiltration equation. Evapotranspiration potential is computed as a function

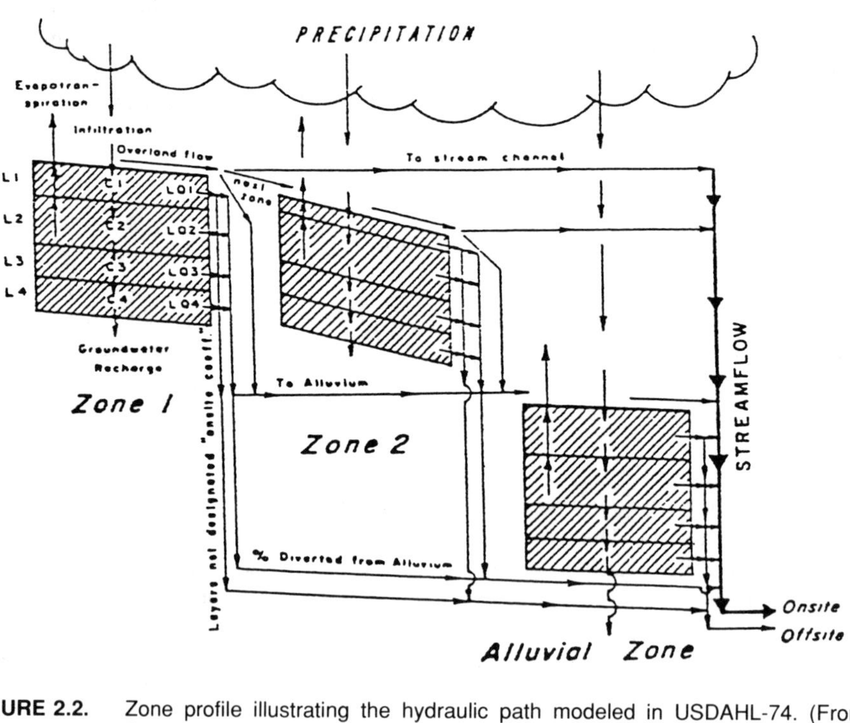

FIGURE 2.2. Zone profile illustrating the hydraulic path modeled in USDAHL-74. (From Fisher et al., 1979. *Trans. ASAE* 22:1345–1352. With permission.)

of crop growth index, pan evaporation, and available soil water. The model maintains a continuous simulation of soil moisture and has a groundwater recharge component. Runoff is assumed to begin as overland flow and to cascade from higher to lower hydrologic zones until it becomes channel flow and is routed downstream.

Climatological input includes breakpoint rainfall, weekly average air temperature, and weekly average of daily pan evaporation. Hydrologic and soil parameter input data include land slope, soil porosity, available water capacity, percentage of surface cracks, final infiltration rate, and the percentage of watershed area occupied by each hydrologic response zone.

For a model to be a useful tool in land use planning, computed streamflow should be sensitive to land use changes. Sensitivity analysis carried out for the USDAHL-74 model by Holtan et al. (1977) showed that there was a good agreement between modeled and recorded streamflow. The model proved to be sensitive to impervious area changes, but good documentation of these changes was required to obtain good results. Because the model was sensitive to some land use changes, Fisher et al. (1979) explored the applicability of the model in watersheds with different physiographic characteristics and investigated the usefulness of the model in evaluating the hydrologic impact of different land use plans. They applied this model to three watersheds under different levels of urban and agricultural development. They found that the model satisfactorily predicted the hydrology of the three watersheds and that the operation of model was not significantly affected by some differences in physiography. Predictions for the mountainous watershed were not as good as the other two, partly because of the snowmelt temperature discrepancy between temperature-measuring stations and the areal average. The thaw temperature had to be adjusted subjectively. Nevertheless, prediction of streamflow for this watershed was fair and, on the whole, this model seems to be a good tool for studying the hydrologic effects of land use changes and thus could be useful in land use planning.

2.2.6 TOPOG

The modeling tools collectively known as TOPOG are being developed in the Centre for Catchment Hydrology of CSIRO, Australia for use in land and water resource management (Vertessy et al., 1990). This model consists of three main parts: (1) a topographic analysis component, (2) a procedure for steady-state analysis of catchment drainage, and (3) facilities for dynamic modeling of catchment hydrologic processes.

The treatment of topography is the main innovation of TOPOG. Landscape under analysis is being segmented into a series of interconnecting hillslope elements that permit flow solutions to be computed in one dimension (O'Loughlin, 1986). For each element derived, a series of attributes such as slope, aspect, and upslope contributing area are calculated and stored. Spatial variation in soil and vegetation properties is accommodated in the analysis by superimposing a series of digitized polygons onto the topography. Having characterized the topography as an element network, simple steady-state analysis is conducted to define the spatial distribution of waterlogging and erosion potential and how these patterns are simulated by different land use strategies. Simulation of the time variable hydrologic behavior of catchments is undergoing development. Test applications to eroded and salinized landscapes demonstrate that the TOPOG package has potential for process modeling of erosion and dryland salinity. Work is currently going on to develop erosion, salinity, and groundwater flow modules for TOPOG.

2.2.7 KINCON

KINCON is a physically based hydrologic model developed to simulate runoff from, and for design of, contour banks and waterway layouts. Infiltration and rainfall excess are calculated using the multilayer transient conductivity Green and Ampt method (Green and Ampt, 1911). Kinematic wave theory is used to route this runoff as overland flow on the interbank plane and in the contour bank channel. Hydrographs from each contour bay are then routed and progressively added down the waterway (Silburn and Loch, 1990).

2.2.8 HILLS

HILLS is a general hillslope simulation model capable of simulating surface runoff and shallow subsurface flow in response to rainfall. The theoretical basis of the model is described by Smith and Hebbert (1983). It is a process-based model that uses well-known mathematical models of individual hydrologic processes to compute the paths taken by rainfall. It is an event model and simulates the response of a hillslope to a single rainfall event, using small time steps appropriate to the time scale of the runoff, infiltration, and subsurface flow processes. Compared to the hydrologic component of ANSWERS (Areal, Nonpoint Source Watershed Environment Response Simulation) (described in Section 2.3.11), this model has the advantage of combining a more complete infiltration model with simulation of a perched aquifer of the type known to occur in lateritic conditions.

The HILLS model requires as input several parameters describing soil conditions that are not readily available. The sensitivity of model predictions to some of these parameters is quite strong. A slight increase in soil porosity, for example, results in a much longer lag time between the rainfall and the peak height of the transient aquifer. This model also exhibits high sensitivity to antecedent moisture conditions.

2.2.9 Model for Hydrologic Delineation of Nonpoint Source Contributing Areas

Most watershed models in use contain surface runoff generation and routing submodels, and generally use lumped input parameters. Models that do accommodate distributed parameter input commonly suffer from the limitation that the spatial cells used in the model may not provide a good fit with hydrologically important features.

The model developed by Gburek (1983) is designed to overcome such limitations. It defines the recurrence interval with which any part of a watershed contributes surface runoff and thus potential nonpoint source pollution to the watershed stream channel system. It is an adaptation of a simple, physically based distributed storm hydrograph-generating model. The original model developed by Engman and Rogowski (1974) incorporated the differing hydrologic responses of various parts of a watershed resulting from the same precipitation input. The portions of the original model incorporated in the new one are the subroutines for partitioning precipitation into infiltration and excess precipitation and overland routing to the channel system. Numerical estimates of the parameters necessary for running the infiltration and overland flow subroutines are available from SCS and survey procedures.

Gburek's model simply delineates the areas of watershed that could provide nonpoint source pollution of the stream via the movement of surface runoff if pollutant existed within this area. It has three subroutines of infiltration/precipitation, surface runoff routing, and maximum travel distance.

Using a series of rainfalls and proper choice of initial soil moisture as input, a map of the watershed can be developed showing the return period with which any part of the watershed contributes surface runoff, and thereby potential nonpoint source pollution, to the stream.

2.3 MODELING SOIL EROSION, SEDIMENT TRANSPORT, AND DEPOSITION

2.3.1 Various Forms of Soil Erosion

Soil erosion may be considered as a natural geomorphic process. It is the acceleration of this process that is of concern. There are a number of erosion mechanisms of importance. There are types of erosion in which flowing water and water as raindrops are directly involved. While these types of erosion will receive almost exclusive attention in this review, there is another broad class of erosion in which water plays only the indirect role of weakening the soil. The direct cause of erosion is the weight of the soil itself. This weight can give rise to stresses in the soil fabric that are too high to be resisted, so that failure occurs, and the resulting erosion is referred to as "mass movement."

Examples of mass movement are landslides and mudflow. However, human action can trigger or accelerate mass movement of soil, examples being the aftermath of clear felling of forests, and instability of a hillside induced by excavation in road construction. In some circumstances cultivation of steep land can lead to the loss of most soil in the cultivated layer during a major rainfall event.

When soil erosion is not due to gravity directly, as in mass movement, but is carried in water moving over the soil surface, there can be more than one process at work. Sometimes an approximately uniform depth of soil is removed in erosion. Rather more commonly water finds some preferential paths on which to flow over the land surface. Then a common, although not universal, characteristic of erosion is for these flow concentrations to deepen the paths in which they flow to form an easily recognizable small channel, referred to as a *rill*. Rills can be formed in a variety of shapes and sizes, and can substantially concentrate overland flow, in the way that rivers do on a much larger scale. Water erosion in which rills are formed is often referred to as *rill erosion*. Rills are ephemeral in the sense that their visual form can be readily removed by cultivation, and in some literature relatively large rills are referred to as *ephemeral gullies*.

The term *gully erosion* is generally used to describe a feature arising from concentrated flow that is too deep to be removed by mechanical cultivation equipment, and may be of such a depth as to be impassable by such equipment. The models of erosion processes and their consequences given below do not cover mass movement or gully erosion, in which mass movement may play a part.

2.3.2 USLE

The most familiar and frequently applied model is the Universal Soil Loss Equation or the USLE (Wischmeier and Smith, 1965, 1978). It has been coded for solution by computer, is used in most SCS field offices, and there is voluminous literature about it and its application.

The USLE uses a conceptual stance common in the early stages of research: the approach is to collect data on all variables thought to be of significance and use general mathematical (statistical) techniques to seek uniformities in them. The outcome of using this methodology on an extensive experimental database obtained in agricultural regions of the U.S. was the well-known USLE product-type relationship:

$$A = RKLSCP \tag{10}$$

where A is the average annual soil loss, R is rainfall erosivity, K is soil erodibility, L is the length of slope, S is the steepness of slope, C is a cropping management factor, and P is an erosion-control practice factor.

A strength of the USLE is the demonstration that Equation (10) can provide a useful summary of a fairly extensive database, adding to confidence that the factors identified in that equation are significant in affecting soil loss. The simplicity of the model and quality of its documentation make it an effective extension technique.

A weakness is that it does not attempt to represent the processes involved in soil erosion, and the database must be completely redeveloped in any new environment so that the parameters in Equation (10) relevant to the new environment can be determined. This limitation applies particularly to factors K, C, and P, although evidence is emerging that the LS factor also is not universal in application.

The model was derived in a period when the conceptual model of soil erosion was that "rainfall detaches and runoff (simply) transports." Hence the emphasis is on rainfall erosivity characteristics [factor R in Equation (10)]. It is now known that runoff itself can act as a potent agent of erosion. Although there will be a connection between rainfall erosivity characteristics (as defined by term R) and runoff, the relationship between them will be site specific. There will be some situations where rainfall detachment is the dominant erosion mechanism, namely at small slopes, and low slope lengths. However, agriculture is not bound to such limitations.

In order to rectify some of these limitations two modifications of the USLE were attempted by Williams (1975b) and by Onstad and Foster (1975), both of which are called MUSLE (Modified Universal Soil Loss Equation). The Williams MUSLE uses an event runoff erosivity factor [11.8 $(VR)^{0.56}$] in place of the rainfall erosivity factor of the USLE:

$$A_e = 11.8\left(Q_e \cdot Q_p\right)^{0.56}\left(\frac{K \cdot LS \cdot C \cdot P}{A}\right) \tag{11}$$

where A_e is event soil loss (tons ha^{-1}), Q_e is event runoff volume (m^3), Q_p is peak runoff rate ($m^3\ s^{-1}$). Other factors of this equation are similar to those used in the USLE. This equation was derived in the United States using soil loss data from catchments up to 2600 km^2. As sediment deposition takes place between the point of entrainment and catchment outlet no delivery ratio was needed. Thus the equation inherently assumes some deposition.

The individual components of the Onstad and Foster MUSLE are similar to the Williams MUSLE. The major difference is that the rainfall-runoff erosivity factor in the Onstad and Foster MUSLE was derived from basic erosion principles. The exact values of the coefficients were determined from rainfall simulator data.

$$A_e = \left[0.646EI_{30} + 0.45\left(Q_e \cdot Q_p\right)^{0.333}\right] K \cdot LS \cdot C \cdot P \qquad (12)$$

where EI_{30} is the erosion index in ton·meters per hectare.

A consensus is emerging that the USLE will probably have a long-term place in soil conservation practice, but that only diminishing returns will be achieved by attempts to add to its development. Hence in recent years attention has shifted to the development of models whose objectives are to represent the processes involved in soil erosion, because it is these processes that are more truly "universal," even though the outcomes can be of great variety. Thus we are seeing a transition or development in objectives in soil erosion research, and a corresponding change in type of model (Rose, 1984).

One reason for attempts, subsequent to development of the USLE, to develop more physically based models is the hope that the parameters defined in such models may have a more definite physical meaning than the *K* factor in the USLE. If so, it then follows that there may be more hope of finding ways in which measurements can be made of soil characteristics, independently of runoff plot experimentation, from which the parameters relevant to erosion can be inferred with an accuracy at least adequate for management decisions needed to reduce soil loss to acceptable levels. Such a development holds out hope (as yet not fully realized) that the assessment of soil erodibility may be made in survey mode, or at least not requiring the installation of instrumented runoff plots to measure soil and water loss, except as a check where this is feasible.

2.3.3 SOILOSS

Determination of local values for the USLE factors and the validation of the USLE model prediction against Australian data have led to the development of a computer program called SOILOSS (Rosewell and Edwards, 1988). This program uses the procedures of the USLE and the available local data to estimate average annual soil loss for New South Wales, Australia.

SOILOSS is designed to make the use of the USLE easier for the everyday user and to facilitate the comparison of erosion control options. The program performs all the necessary calculations and only requires information relating to the location, soil type, topography, land use, and crop management at a site. The required information is either selected from the guide, from the menu in the program, or determined by

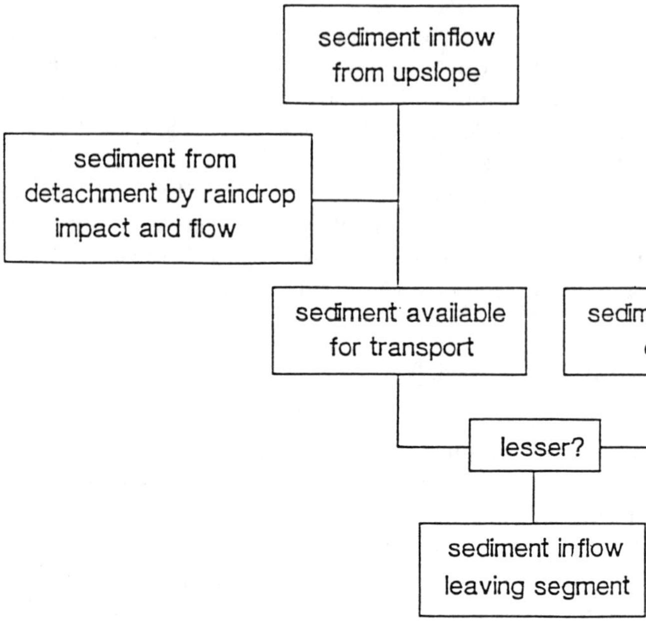

FIGURE 2.3. Conceptual model of interaction of erosion and transport processes. [From Foster, 1980. In: Environmental Impact of Nonpoint Source Pollution. M. R. Overcash and J. M. Davidson (Eds.). Ann Arbor, MI: Ann Arbor Science, pp. 213–243. With permission.]

measurements on the site in question. SOILOSS contains algorithms to estimate the soil erodibility factor from soil data and uses crop growth, tillage, and residue management models to provide an estimate of the crop management factor.

2.3.4 The Soil Erosion Model of Foster

This model divides the erosion process, on the basis of morphology, into rill processes and interrill processes. Interrill processes are made dependent on the rainfall rate or on terms that have their origin in the USLE. The basic equation describes the spatial increase in sediment flux as being the sum of an interrill contribution together with a rill erosion term. The potential sediment load for rill transport is compared with an upper limit to transport (the "sediment transport capacity"), which is determined using a modified Yalin bed-load transport equation. Net deposition is assumed to occur if the transport capacity is exceeded, otherwise the transport is limited to the estimated potential load (Figure 2.3). The description of fluxes in these two components is not geometrically explicit, despite the conceptual separation into rill and interrill components, so detail on rill geometry or frequency is not needed (Foster, 1980).

There is evidence that rilling increases the efficiency of sediment transport when it occurs, although the prevalence of rilling is not universal, depending among other things on soil type and condition. Where rill frequency is high it is useful and reasonably correct to identify the erosion mechanism in the limited interrill regions as rainfall detachment, with entrainment (or "runoff detachment") dominating in rills and this occurs in the relatively high-silt soils common in some major agricultural regions of the U.S. The identity between surface morphology and erosion process is very widely used (Foster, 1982). However, this identity is not universal. Rilling is much less common or virtually absent in some soil types and conditions, especially in tropical soils. The lack of rills does not imply absence of the entrainment processes, which dominate in rills. Erosion between (or without) rills can also arise from a variety of processes, and the term "interrill erosion" does not describe a unique process.

Thus, to avoid confusion and allow generality it is recommended that the names of the erosion mechanisms be used in describing erosion processes, rather than the implied process descriptors *rill erosion* and *interrill erosion*. This does not deny the utility of the terms *rill* and *interrill regions* in describing a common erosion surface morphology.

2.3.5 WEPP

An alternative approach to the USLE is being developed by a group of USDA scientists from a number of locations in the U.S. including the National Soil Erosion Laboratory of the Agricultural Research Service of the USDA (USDA-ARS), the birth place of the USLE. The Water Erosion Prediction Project, or WEPP, seeks to improve erosion prediction technology, and is to be based on the hydrology of surface flow and water erosion processes. It uses the methodology of Foster (1987). It is modular so that new research findings can be incorporated without major revision of the model. The first version of WEPP has been delivered and is being evaluated and tested by user groups while research continues to develop and refine relationships in the model. The version for wider use is expected to be delivered in 1992. The WEPP model will have three versions: the already available hillslope version that applies to landscape profiles similar to those for the USLE except that the WEPP model considers deposition areas; a small watershed version using a representative landscape profile and considering waterways within the application area; and a grid version to compute sediment movement at all points in all waterways over a catchment or watershed. As shown in the model flow chart (Figure 2.4), the WEPP model is constructed to describe the influence of climate, soil, topography, cover management, and supporting practices on erosion, deposition, sediment yield, and sediment characteristics. The technology incorporated within this model is expected to have a life span of about 20 years (Foster, 1987; Lane et al., 1988; Lane and Nearing, 1989).

Some of the main erosion features of the WEPP model are as follows:

1. The erosion component is divided into hillslope and channel erosion. The hillslope erosion process is in turn divided into interrill (a function of soil and rainfall intensity) and rill (a function of soil, flow shear stress, and transport capacity) components. Locations of detachment and deposition on the hillslope are calculated for each erosion event. Hillslope erosion equations used in the WEPP model are as follows:

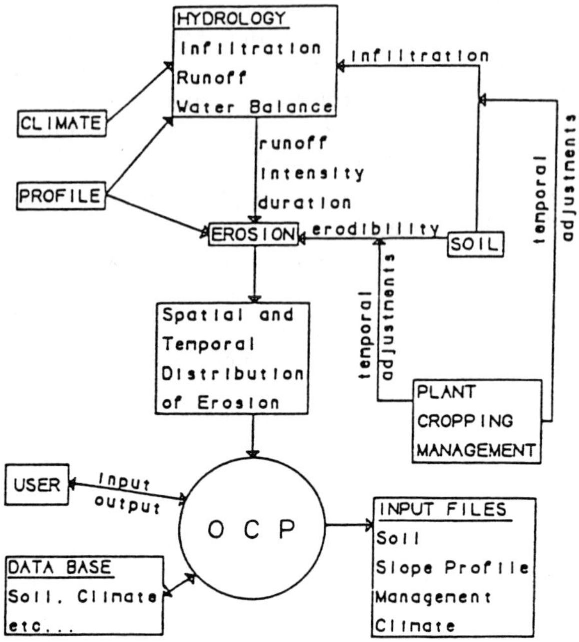

FIGURE 2.4. Flow chart for the WEPP model. (From Lane and Nearing, 1989.)

Erosion equation $$dM / dx = D_r + D_i \tag{13}$$

Rill erosion $$D_r = K_r(\tau_f - \tau_c)(1 - M / T_c) \tag{14}$$

Interrill erosion $$D_i = K_i L_e^2 C_e G_e (R_s / w) \tag{15}$$

Sediment yield $$M = M^* T_{ce} (w / R_s) \tag{16}$$

where M is sediment load (kg s^{-1} m^{-1}), x is distance downslope (m), D_r is rill erosion rate (kg s^{-1} m^{-2}), D_i is interrill erosion rate (kg s^{-1} m^{-2}), K_r is rill erodibility coefficient (s m^{-1}), τ_f is flow shear stress (Pa), τ_c is critical shear stress (Pa), T_c is sediment transport capacity in the rill (kg s^{-1} m^{-1}), K_i is interrill erodibility coefficient, L_e is effective rainfall intensity (ms^{-1}), M^* is sediment load normalized to transport capacity, C_e is interrill canopy effect coefficient, R_s is rill spacing (distance between rills) (m), G_e is the effect of ground cover on interrill erosion, w is computed rill width (m), and T_{ce} is transport capacity at the end of uniform slope (kg s^{-1} m^{-1}).

The channel erosion component computes the detachment, deposition, and transport of sediment within a channel. Locations of detachment and deposition along a channel are calculated. The channel erosion equation is

$$dM / dx = D_{cf} + q_1 \tag{17}$$

where D_{cf} is net detachment or deposition rate (kg s^{-1} m^{-2}) and q_l is lateral inflow sediment load to channel (kg s^{-1} m^{-2}).

2. The hydrologic component consists of a climate generator (based on CLIGEN model), rainfall disaggregation (a time intensity pattern of daily rainfall), Green and Ampt infiltration, peak discharge (semianalytical solution of the kinematic wave equations), and spatially varied flow equation.

2.3.6 Water Erosion Model of Rose and Hairsine (GUEST)

The model of Rose and Hairsine (1988) brings together an approximate analytical model of overland flow (Rose et al., 1983b), and a description of erosion processes and deposition in the context of an equation describing the conservation of mass of sediment. The processes affecting sediment concentration are conceived as acting continuously and therefore simultaneously. The processes as described in the model are given below. More detailed comment on this approach is given because of author involvement in developing this model.

2.3.6.1 Deposition

The majority of sediment eroded from the soil surface will return again to the soil surface after being carried some distance downslope in overland flow. This process, whereby sediment falls through a water layer, is called "deposition". The characteristic of soil that governs the rate of deposition is called the "settling velocity characteristic". This characteristic of soil can be readily measured experimentally by determining the proportion of soil mass with any particular settling velocity. The characteristic can be measured with rather simple equipment called a modified bottom withdrawal tube described by Lovell and Rose (1988), who also give the theory of operation of the equipment. A computer program is helpful in analyzing results, as the calculations are tedious (Lovell and Rose, 1986).

Figure 2.5 illustrates settling velocity characteristics determined using the modified bottom withdrawal tube presented as a cumulated percentage of sediment that settles with a velocity slower than any settling velocity. The figure shows that no sediment in this soil settled with a velocity faster than somewhere between 0.1 and 0.2 m s^{-1}.

In erosion theory for equilibrium or near-equilibrium conditions (Rose and Hairsine, 1988), it is shown that a single quantity derived from the settling velocity characteristic represents the role played by deposition in controlling sediment concentration. This single quantity, called the "depositability" of soil, is calculated by reading off the settling velocity at regular steps of the ordinate in Figure 2.5, and then finding the mean of the sum of such settling velocities. The depositability is defined as $\Sigma\, v_i / I_t$, where I_t is the total number of equal mass intervals into which the ordinate in Figure 2.5 is arbitrarily divided.

Let c_i denote the sediment concentration of the class of particles (or aggregates) whose settling velocity is v_i. Then it follows from the definition of these terms that the rate of deposition (d_i, defined as the mass of sediment reaching the soil surface per unit area per second) is given by:

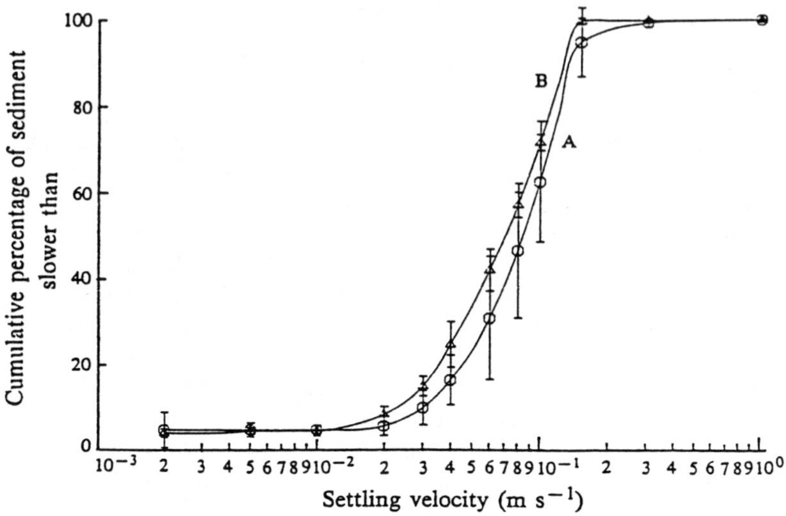

FIGURE 2.5. Settling velocity distributions for sediment of black earth deposited during a field erosion event and analyzed while still moist (curve A); for the same sediment allowed to air dry and rewet by immersion (curve B).

$$d_i = v_i c_i \quad \left(\text{kg m}^{-2}\ \text{s}^{-1}\right) \tag{18}$$

Sediment concentration depends on the difference between the rate of sediment addition to overland flow by the active erosion processes and the rate of deposition, given by the summation of d_i in Equation (18) over the total velocity range of depositing sediment. Erosion processes, and factors on which their rates depend, will be considered in the following sections.

2.3.6.2 Rainfall Detachment and Redetachment

The removal of soil for the first time in an erosion event by the impact of a raindrop is referred to as rainfall detachment. Such detachment is caused by the locally intense shear stresses generated at the soil surface unless it is protected by a significant depth of water or by other cover. In time some of the detached sediment settles back to the soil surface (or deposits), covering at any time some fraction (H) of the original soil surface. This layer of deposited material is also subject to rainfall impact, and its removal by rainfall impact is referred to as "redetachment" (Rose and Hairsine, 1988).

A major reason for distinguishing between detachment of original soil and redetachment of the deposited layer lies in the strength of the material being detached, which can have a major effect on its rate of removal. The original soil will typically possess some cohesive strength. Previously detached sediment deposited on the soil surface typically has little time to rebuild cohesive links with neighboring particles, and thus the deposited layer is very weak (Rose et al., 1990).

Rainfall detachment and redetachment can be the only erosion processes on relatively flat regions of small extent. For such small, low-slope regions any slow-moving overland flow plays only the passive role of simply transporting sediment whose concentration is determined by the net outcome of rainfall detachment and redetachment adding sediment to the water layer, and deposition removing sediment.

2.3.6.3 *Rainfall Detachability and Redetachability*

The determination of depositability was described first, because theory outlined earlier indicates that the types of erodibility parameters defined in this theory require, for their evaluation, a knowledge of depositability ($\Sigma\, v_i / I_t$).

The theory of sediment concentration and transport at equilibrium or near-equilibrium conditions in the absence of any flow-driven processes is given by Hairsine and Rose (1991). This theory defines as one erodibility parameter the detachability of the original soil matrix to rainfall impact. This detachability (a) is defined by

$$e = aR_r \quad \left(\mathrm{kg\ m^{-2}\ s^{-1}}\right) \tag{19}$$

where e is the rate of detachment per unit area of soil by rainfall of rate R_r. The value of a is at maximum for shallow water depths, but decreases for greater depths.

However, after a time much of the soil surface becomes covered by a deposited layer. At equilibrium this coverage is so relatively complete (>90%) that more important than the soil detachability given in Equation (19) is the redetachability (a_d) defined in relation to equilibrium sediment concentration in Equation (20):

$$c = \frac{Ha_d R_r}{\sum v_i / I_t} \quad \left(\mathrm{kg\ m^{-3}}\right) \tag{20}$$

where H, the fractional coverage of the soil surface by the deposited layer, appears to have a value of approximately 0.9 at equilibrium (Proffitt et al., 1991), a_d is a redetachability coefficient, I_t is an arbitrary number of settling velocity classes to which sediment is divided, and R_r is rainfall rate. At equilibrium, the detachability of the original soil plays a relatively minor role in affecting the sediment concentration, although it affects the time required to achieve an equilibrium situation.

The value of a_d can be calculated using Equation (20) from measured c and R_r (assuming v_i/I_t known from the settling velocity characteristics, and $H = 0.9$). Theory (Hairsine and Rose, 1991) then shows that

$$a = Q / \left[R_r / c\left\{\left(\Sigma v_i / I_t\right) / a_d\right\}\right] \quad \left(\mathrm{kgm^{-3}}\right) \tag{21}$$

where Q is runoff rate per unit area.

2.3.6.4 *Runoff Entrainment and Reentrainment*

When water flows over a sloping land surface, a shear stress develops between the land surface and the flowing water. At equilibrium or nonaccelerating flow, the stress acting upslope on the flow must be equal and opposite to the component of the weight of water acting down the plane. Also, the stress acting upslope on the water has an equal and opposite reactive stress acting downslope on the soil. Removal of original soil from the surface due to this stress can be called "entrainment". (Terminology is not uniform, and the term "runoff detachment" is also in use to describe this process.)

In considering rainfall detachment and redetachment, we saw that erosion altered the nature of the soil surface by partly covering the original soil surface by a deposited layer of sediment, generally weaker than the original soil. This partially blanketing deposited layer also quickly forms when runoff entrainment is the erosion process. Sediment is removed from this deposited layer due to exactly the same mutual shear stresses acting between the soil and overland flow as caused by entrainment, and such removal is referred to as "reentrainment".

Distinction between entrainment and reentrainment is important because of the weaker nature of deposited sediment. A significant amount of work must be done in entraining original soil and tearing it loose from the cohesive forces holding it to its neighboring soil. In contrast, the deposited sediment does not have time to reestablish cohesive bonds before it is removed again.

Bagnold (1977) introduced the concept of "stream power" (Ω), defined as the rate of work of the shear stress (τ). Thus, if V is the velocity of flow, then

$$\Omega = \tau V \quad \left(W\ m^{-2}\right) \tag{22}$$

This stream power is the source of energy for both entrainment and reentrainment. The theory for the processes is given in Rose and Hairsine (1988). It assumes that not all the energy expended is used in entraining or reentraining sediment, but rather that only some fraction F is effective in such erosive processes. The value of F must be somewhere between zero and one, and be determined experimentally.

Work carried out in the Griffith University Tilting Flume Simulated Rainfall facility has found that $F \approx 0.1$ for turbulent flow. This value has been shown (Proffitt, 1988) not to depend on soil type, and to increase to about 0.2 for laminar flow. Thus, fraction F appears to be a characteristic of fluid behavior, and does not depend, for example, on whether either entrainment or reentrainment is the dominant erosion process, or the type of soil being eroded. The rate of working of the shear stress effective in eroding soil is thus $F\Omega_0$, and the remaining 90% of stream power [or $(1 - F)\Omega$] ends up as heat.

It is assumed that the maximum sediment concentration will be achieved if the soil surface is completely covered by a layer of freshly deposited sediment (i.e., $H = 1$). Then entrainment will not occur and reentrainment is the only erosion process. It is also assumed that freshly deposited sediment is of negligible strength. Thus, the work required to reentrain sediment is only that needed to lift the sediment against its immersed weight in water (whereas in entrainment work would also be done against cohesive forces in removing sediment from neighboring sediment).

Foster (1982) drew particular attention to the experimental finding that there is an upper limit to the sediment concentration even in situations where entrainment and reentrainment are dominant, and introduced the term "transport limit" to describe this. This term is widely used in the soil erosion literature and, denoting it by c_t, the theory developed in Rose and Hairsine (1988) shows that

$$c_t = \frac{F\rho}{\sum v_i / I_t}\left(\frac{\sigma}{\sigma-\rho}\right)SV \quad (\mathrm{kg\ m^{-3}}) \tag{23}$$

where ρ is the density of water, σ is sediment density, S is slope, and V is mean flow velocity of water.

The derivation of Equation (23) assumes the soil surface to be completely covered by previously entrained and deposited material, assumed to be effectively cohesionless. Complete coverage by a deposited layer (i.e., $H = 1$) is unlikely to be sustained for any substantial time period, since with $H = 1$ the only source of sediment is the deposited layer itself. Once the deposited layer becomes less than complete in its coverage, then the original cohesive soil is exposed, and some of the stream power is expended in entraining such cohesive sediment.

2.3.6.5 Theory for Entrainment and Reentrainment Acting Together

Since all soils except pure sands usually exhibit some cohesive strength, there must be some energy expended in entraining any mass of cohesive soil. Rose and Hairsine (1988) introduced the fundamental erodibility parameter (J), which is the energy per unit mass of soil entrained, or the specific energy of entrainment. The energy expended in entraining unit mass of soil must be greater than the energy used to simply lift sediment against its immersed weight. Thus, the sediment concentration (c) measured when entrainment, as well as reentrainment, is taking place must be less than the upper limiting value c_t given by Equation (23). Thus, we can write, in general terms,

$$c = c_t - f(J) \quad (\mathrm{kg\ m^{-3}}) \tag{24}$$

where $f(J)$ is a function of J, which must be experimentally determined.

Experimental results of Rose et al. (1990) illustrate the likelihood that there is a lower as well as an upper limit to sediment concentration. Theory can be developed (Rose and Hairsine, 1988; Hairsine and Rose, 1991) that indicates a basis for such a lower limit, set by the value of the erodibility parameter J defined above (and other known parameters). This lower limit depends on the value of J for the source of sediment, which is the original soil matrix, and this limit has therefore been called the "source limit".

Using Equation (24), if c is measured, c_t calculated from Equation (23), and the form of $f(J)$ is known from theory, then the specific energy of entrainment (J) can be calculated (Rose et al., 1990). Theory for the function $f(J)$ in Equation (24) can be derived by considering the equation for mass conservation of sediment. For situations close to equilibrium, the time rate of increase of storage of sediment can be neglected relative to the spatial rate of increase of sediment transport. Solution of the ordinary differential equation involving J requires numerical solution of a differential equation, and a program called GUEST (Griffith University Erosion System Template) (Misra and Rose, 1989) is available to aid such evaluation.

However, a relatively simple analytical solution in algebraic form of the differential equation involved can be obtained, assuming the value of 2 for the parameter m, in the kinematic flow approximation to describe the volumetric flux (q) on a plane:

$$q = K_c D^m \quad (\mathrm{m}^2\ \mathrm{s}^{-1}) \tag{25}$$

where K_c is dependent on slope and roughness of the plane surface over which the flow is occurring, and D is the depth of flow. Taking $m = 2$ indicates a state of flow somewhat closer to turbulent conditions (for which $m = 1.67$) than to laminar conditions (when $m = 3$). The other assumption is that overland flow occurs on a planar land surface without rills. For $m = 2$, the function $f(J)$ in Equation (24) is given by Rose and Hairsine (1988).

The solution is illustrated in Figure 2.6 for assumed values of J. The value of other parameters are taken from Proffitt (1988) for a vertisol (or black earth): $\Sigma\, v_i / I_t = 0.06$ m s^{-1}, $\sigma/(\sigma - \rho) = 2$. Assuming K_c in Equation (25) is given by $K_c = S^{1/2}/n$, then Manning's $n = 0.03$ and $S = 0.05$ are assumed. Also the runoff rate per unit area, Q, is taken as 50 mm h^{-1} (or 1.39×10^{-5} m s^{-1}).

Figure 2.6 shows c plotted as a function of stream power (Ω), first for the transport limit, when $c = c_t$ and $J = 0$, but also for increasing values of J. Since entrainment and reentrainment are the only erosion processes represented, c is shown as falling to zero for $\Omega = 0$, whereas there could be a significant contribution to c from rainfall for all Ω, including $\Omega = 0$.

The same concepts as described for flow over a plane surface also hold when rilling occurs, the increased complexity in equations arising from the more complex geometry involved with rilling. Because of the availability of computer programs (Misra and Rose, 1989), this complexity is hidden in their use. This analysis provides

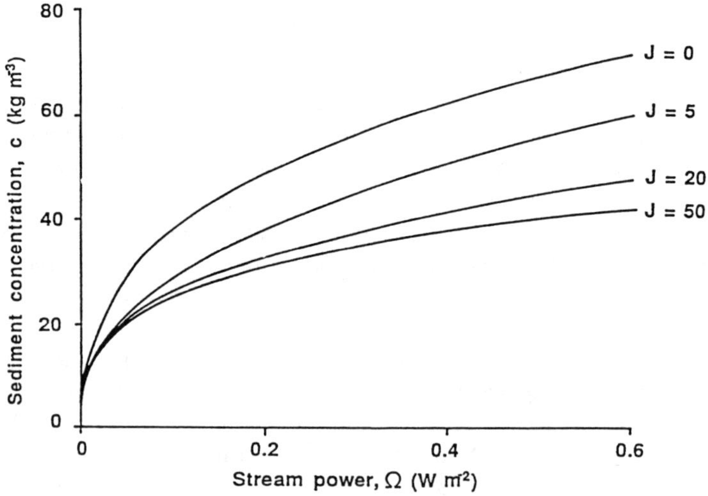

FIGURE 2.6. Variation with stream power in sediment concentration at the transport limit ($J = 0$), for particular assumptions regarding soil, slope, and runoff characteristics. Also shown is the variation in sediment concentration at the entrainment limit assuming the values shown for the parameter J. (From Rose et al., 1990.)

a theoretical interpretation of the well-known fact that rilling increases sediment concentration when it occurs, other factors being equal or similar (e.g., Proffitt et al., 1991). The basic reason for this increase is that channeling flow into rills considerably increases the (local) stream power of flow in rills compared to the same overall flow rate per unit area in uniform flow.

Following cultivation, soil can gradually regain strength, leading to a decline in sediment concentration, as well as a decline in dependence on land slope and slope length.

2.3.6.6 Erosion and Deposition Processes Acting Together

Each process has so far been considered in isolation. Of course all processes commonly act together at the same time. Thus both the rainfall and runoff-driven erosion processes add to sediment concentration, with deposition reducing it. The ability to imagine all the processes acting simultaneously is illustrated schematically in Figure 2.7, where the rates are shown as valve symbols, and the sediment is shown artificially elevated above the soil surface to allow the rate processes to be shown more clearly.

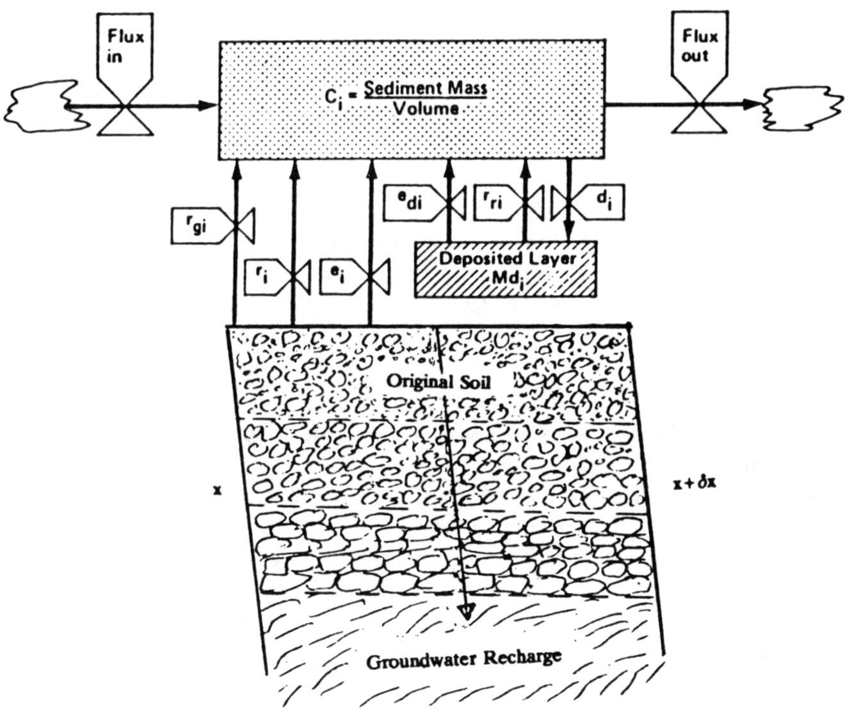

FIGURE 2.7. Flow diagram describing the erosion and deposition processes in the model of Rose and Hairsine. Rates e_i and e_{di} refer to rates of rainfall detachment and redetachment for size class i; r_i and r_{ri} refer to rates of entrainment and re-entrainment; r_{gi} refers to gravitational sediment flux density. (From Proffitt et al., 1989.)

It is the outcome of all these processes acting together that is of practical concern. However, because the individual processes depend on quite different soil characteristics or external factors, penalties are paid if the processes are not separately described.

The relative importance of rainfall detachment and redetachment in comparison to runoff and entrainment has been shown by Proffitt (1988) to depend on the magnitude of the stream power of overland flow. As stream power increases, the importance of rainfall impact as an erosive process diminishes relative to those associated with overland flow.

2.3.6.7 Gravity-Driven Processes and Mass Movement of Soil

If the transport limit is to be sustained, it is clear that entrainment cannot be the source of this previously eroded sediment now forming a deposited layer completely blanketing the original soil matrix. What must be the type of process or processes that can give rise to the sediment forming this complete deposited layer?

In situations of substantial erosion rainfall detachment is unlikely to be more than a very minor contributor to such sediment. Thus the only possible answer to this question of the source of a layer of deposited sediment that is complete at least in the bottom of rills must lie in movement of the soil mass by gravity, even though other processes may play a part. The common observation of rill wall collapse, sometimes stimulated by bank undercutting or resulting from rill wall strength failure, is one such gravity-driven process. Rill head cutting is another.

Thus if sediment concentration is to achieve the transport limiting value, and be sustained at this limit, then there must be a rate of addition of sediment due to processes deriving their energy directly from gravity. Such a rate process is shown schematically on the top of Figure 2.7. It is likely that such processes as rill wall collapse and head cutting will be stochastic in nature, but may occur with sufficient frequency for the deposited layer to completely blanket the original soil matrix, and so for the transport limit concentration to be sustained. Such gravity-aided processes are expected to play an even greater role in more extreme forms of erosion that lead to gullies.

2.3.7 Erosion Model of Yang and Its Application by Moore and Burch

Yang (1973) introduced the concept of "unit stream power", defined as the time rate of dissipation of the potential energy per unit weight of water involved in overland flow. From an extensive series of experiments on noncohesive stream-bed materials in flumes Yang (1973) found that the measured sediment concentration could be described by an empirical relationship involving unit stream power, and that the form of this relationship appeared to have some generality.

Because of the noncohesive nature of the materials used in Yang's experiments, it may be assumed that the equation describes sediment transport in a "transport-limiting" or "transport-capacity" situation, with sediment of zero strength. Moore and Burch (1986) used Yang's relationship to describe sediment transport capacity, and showed how it could be usefully applied in a range of situations, including both sheet and rill flow. For both shallow overland flow and rill initiation they found that incipient motion of sediment could be described in terms of a critical unit stream power independent of slope. They also investigated the effect of rill geometry on flow in rills, illustrating the utility of a "rill shape factor" in hydraulic description.

Moore and Burch (1986) showed that the empirically based length-slope factor in the USLE could be provided with some degree of theoretical support in terms of unit stream power concepts if it is assumed that soil loss in many of the erosion experiments in the USLE database were transport limited.

Burch et al. (1986) combined the above approach to sediment transport capacity with the three-dimensional or topographic analysis of O'Loughlin (1986), and showed that the sediment transport capacity at all points within a three-dimensional catchment could be estimated by making simplifying assumptions. In this way the relative erosion hazard or intensity of net deposition likely in different zones in a catchment could be estimated, providing a practically useful basis for conservation planning at a catchment scale. Application of this model to a small experimental catchment at Wagga Wagga in Australia showed good spatial agreement between predicted zones of severe sheet erosion and observed gullies (Burch et al., 1986).

2.3.8 IUSG

The sediment transported in a stream consists of suspended load and bed load. If the bed-load contribution to the total sediment is assumed negligible, then suspended sediment yields reflect watershed sediment yield processes. Conceptual models for computing sediment graphs have been reported by Renard and Laursen (1975), Bruce et al. (1975), and Williams (1978a,b) but a large number of parameters involved has made their application difficult.

Williams (1978b) has developed a model for estimating the sediment graph for ungauged watersheds, given the total sediment yield. Storm sediment graphs are predicted by convolving rainfall excess with an instantaneous unit sediment graph (IUSG). Use of the convolution (or the superposition integral) assumes the system to be linear, which assumption may not be satisfied by sediment flow.

IUSG is the time distribution of sediment resulting from an instantaneous burst of rainfall producing one unit of mobilized sediment. The conceptual model, developed by Kumar and Rastogi (1987), was obtained by routing the mobilized sediment through a cascade of equal reservoirs in which sediment storage is assumed proportional to sediment discharge. It does not account for the geometric configuration of any given watershed (Kumar and Rastogi, 1987) but fulfills a similar function to be unit hydrograph in hydrology. The suspended sediment dynamics is represented by a spatially lumped continuity equation and a linear storage law. Measured and computed sediment graphs show that the convolved sediment graphs are good approximations of the actual storm sediment graphs. The model generates a sediment graph of similar form to that observed (Kumar and Rastogi, 1987), although a rather large relative error may occur at both ends of it.

Kumar and Rastogi (1987) showed how analysis of basin IUSGs could be used to assess trends in the suspended sediment yield of an Indian watershed that had experienced changes in land use and management designed to conserve soil.

2.3.9 GAMES

The University of Guelph model for evaluating the effects of Agricultural Management Systems on Erosion and Sedimentation (GAMES) was developed to identify problem areas on agricultural watersheds and to evaluate the effect of alternative solutions on nonpoint source pollution (Cook et al., 1986; Dickinson et al., 1984). An attempt has been made to keep the model simple, applicable to fields of any size and shape, and using available input data. It describes the significant processes seasonally, and accommodates changes in soil and crop management conditions. GAMES requires a relatively limited amount of readily available input data, and in this respect has advantages over other distributed models. It provides a way of studying spatial and temporal variabilities and the distribution of erosion and sediment problem areas as well as the relative impacts of alternative remedial strategies.

The major components of the model are a soil loss component, fundamentally the USLE, and a sediment delivery component. The sediment yielded by each field to downstream fields and streams is determined from

$$A_s = \mathrm{DR} \times E_s \tag{26}$$

where A_s is the seasonal sediment yield delivered from one field to another or to a stream (tons ha^{-1}), DR is the seasonal delivery ratio between two selected points (i.e., from field to field or from field to stream), and E_s is the product of the soil loss per unit area (determined by USLE) and the field area. DR is determined as per Dickinson et al. (1986a),

$$\mathrm{DR} = f\left(V_s / L_d\right) \tag{27}$$

or

$$\mathrm{DR} = \alpha[\Sigma^{0.5} H_c / nL_d]^{\beta} \tag{28}$$

where V_a is the average velocity of overland flow between locations, L_d is the length of surface drainage path between locations, n is Manning's surface roughness coefficient, S is the slope gradient of the soil surface, H_c is a hydraulic coefficient introduced to represent the capability of a location to generate surface runoff in a selected season and α and β are constants. The possible effects of runoff entering the area from upslope are neglected.

This expression was calibrated on the basis of erosion estimates, time frame selected, and the sediment load measured at the outlet on a small watershed (Dickinson et al., 1986a).The watershed was divided into land cells, each with single land use, single soil type, and single slope class.

Gross erosion for each cell (for each season) was calculated using the USLE. Results suggest that the microdelivery ratios across a watershed can vary significantly as a result of variations in slope and surface roughness values. Dickinson et al. (1986a) found DR to vary over the range of 0.08 to 0.24. On the whole the variability of the delivery ratio is most pronounced in steep slope/low roughness zones. The microdelivery ratio expression is capable of estimating the proportion of sediment transported from field size areas to adjacent areas and to the stream and outlet of a small agricultural watershed for a seasonal time frame, but only following calibration.

One of the limitations of this expression is the assumption of no change in physical characteristics of entrained sediment in the watershed during the erosion process. According to Rose's settling velocity theory and the size selectivity of deposition process (Rose, 1985), it is unlikely that physical characteristics of sediment remain unchanged. The second problem with this expression is the need to obtain parameters by calibration, and therefore like all empirical models it requires experimental data for sediment delivery that are not always available. The potential advantage of the models based on physical principles is that even though the value of defined local variables may still require measurement, this measurement can be determined independently of the event to be described.

Dickinson et al. (1986b) applied GAMES to several watersheds in southern Ontario and found good agreement between the measured and GAMES-predicted soil erosion, delivery ratio of sediment delivered to the stream, and sediment yield per hectare. The procedure divides watershed into four types of problem areas: areas exhibiting both erosion and sedimentation problems, areas exhibiting only erosion problems, areas with only sediment problems, and areas with no problems. This approach provides a logical tool for watershed soil erosion and sediment control planning. The basic strength of the model is due to its distributed nature, which takes into account temporal and spatial variation in relevant factors and identifies erosion and sediment hot spots in a watershed. Although evaluation of model accuracy is very difficult, the model is useful for planning and implementation of effective and economical ways of studying remedial control strategies. It also facilitates consideration of the best management practices for different areas of a watershed (Dickinson et al. 1986b).

2.3.10 KYERMO

The Kentucky Erosion Model (KYERMO) is a physically based research model developed by Hirschi and Barfield (1988a). It is intended to be used to estimate sediment loss on steep slopes. KYERMO utilizes the Meyer and Wischmeier (1969) conceptual model for erosion process with the plot divided into slope increments and rill watersheds on each increment. Unlike the Meyer and Wischmeier model the sediment transport and detachment rates are calculated in individual rills. The main program, shown in Figure 2.8, has four components: runoff generation, runoff routing, sediment generation, and sediment routing, each with a number of subroutines and functions.

The runoff generation component consists of surface storage, infiltration, and runoff calculation routines. To estimate total available surface storage, curves developed by Linden (1979) are used. These predict zero storage for slopes above 20%. The infiltration subroutine is based upon work by Moore (1981) and Moore and Eigel (1981), which are an extension of the Green-Ampt-Mein-Larson (GAML) equation developed by Mein and Larson (1971).

Moore and Eigel (1981) modified the GAML model to account for a two-layer system that in turn was modified for field saturation in KYERMO. The runoff

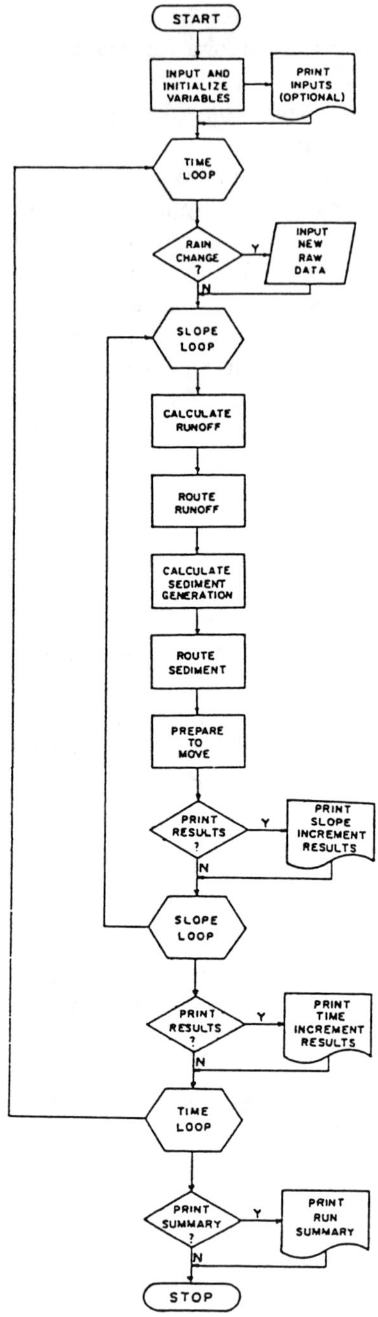

FIGURE 2.8. KYERMO flow chart. (From Hirschi and Barfield, 1988a.)

subroutine accounts for surface storage using a modification of an exponential relationship proposed by Tholin and Keifer (1960), allowing runoff to occur while surface storage is only partially filled. It is assumed that rainfall that does not become runoff enters the surface storage. This equation gives the correct trend but does not converge to runoff equaling rainfall minus infiltration when the potential surface storage is filled. It only does so as ($R_u - I_v$) gets very large, where P and F are cumulative rainfall and infiltrated volumes, respectively.

The runoff routing component of the model routes runoff through the dynamic rill network where cross-sections are changed with time in response to erosion of sediment. Flow in each rill is calculated on a slope increment using Manning's equation.

In the sediment generation component, the processes of raindrop detachment, rill flow removals, rill wall slumping, and change of rill cross-section caused by erosion are calculated and used for the calculation of total sediment detached or entrained and available for transport downslope. This value is then used in the sediment routing component of the program.

In routing sediment the transport and deposition of sediment is calculated separately for sheet and rill flow. Sediment transport capacity of rill or sheet flow is calculated either by using the Foster and Meyer (1972) modification of the Yalin (1963) equation, or by a modification of the Yang (1973) equation.

A sensitivity test carried out by Hirschi and Barfield (1988b) suggests that for steep slope flow the detachment capacity, rather than the sediment transport capacity, is the limiting factor for sediment delivery. The transport capacity becomes the limiting factor as slopes become milder. It was also shown that detachment coefficients, critical tractive force, number of rill per unit width, number of slope increment, and Manning's n are important parameters in sediment yield. Parameter estimation procedures must be found if accurate results are to be expected from this model, and the lack of such procedures is the major limitation of the model for determining absolute erosion amounts. A prior knowledge of rill density is a requirement for KYERMO usage. The erosion prediction of KYERMO using parameter values available in the literature is as good as any other model, such as CREAMS. Simulation accuracy can be increased by parameter value optimization.

2.3.11 ANSWERS

The Areal, Nonpoint Source Watershed Environment Response Simulation, or ANSWERS, is a distributed parameter, deterministic, and primarily event-oriented model developed for predicting the hydrologic response of watersheds to storms and erosion and sediment response of watersheds to different agricultural management systems (Beasley and Huggins, 1980; Beasley et al., 1980). It is designed as a planning tool for those concerned with nonpoint source pollution originating from agricultural lands. It can identify critical areas within a watershed that have high erosion rates, and determine whether or not the soil losses from these areas contribute substantially to the total sediment yield of the watershed. ANSWERS essentially consists of a hydrologic and an erosion model that have been interfaced to provide hydrologic and erosion responses from given data. The conceptual basis for the hydrologic was developed by Huggins and Monke (1966). The governing equation for the simulation of erosion is the continuity equation used by Foster and Meyer (1972).

In using ANSWERS, watersheds are conceptually divided into squares of uniform characteristics as illustrated in Figure 2.9. Within each element, the runoff and erosion processes are treated as independent functions of hydrologic and erosion-related parameters of that element. The hydrologic response of each element to water input is computed by solving the continuity equation. The outputs of each element become sources of input to adjacent elements. The water flow component is provided by Manning's equation for overland flow. Overland and tile-drain outflow from one unit flows into the neighboring unit according to the direction of slope until it reaches a channel element. Overland flow initiates when surface retention is exhausted and the residual is greater than infiltration.

The governing equation for the erosion model is the continuity equation of Foster and Meyer (1972) described earlier [Equation (17)]. Detachment of soil particles by raindrop impact is calculated using the relationship described by Meyer and Wischmeier (1969):

$$R_d = 0.027\ C\ K\ A_i\ R_r^2 \tag{29}$$

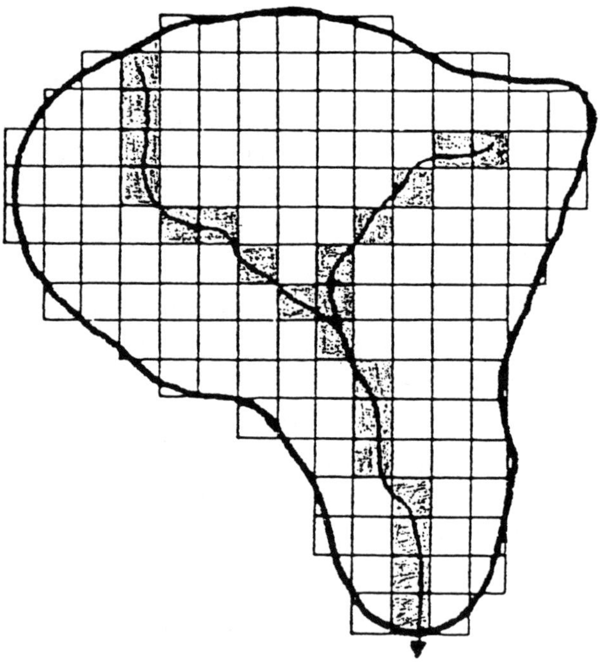

FIGURE 2.9. Watershed divided into elements with channel elements shaded. (From Beasley et al., 1980. *Trans. ASAE* 23:938–944. With permission.)

where R_d is rainfall impact detachment rate (kg min^{-1}), C is cropping and management factor of the USLE, K is soil erodibility factor, A_i is area increment (m^2), and R_r is rainfall intensity (mm min^{-1}).

Land use changes, tillage techniques, and management procedures are simulated by using appropriate values.

The public version of ANSWERS can only model water and sediment yield from the watershed. This model does not require extensive calibration. A research version of ANSWERS can predict the particle size distribution of eroded sediments and the transport of individual particle size classes (Dillaha and Beasley, 1983). This ability allows the assessment of the clay enrichment of the outflowing sediment (Beasley et al., 1982).

The primary strength of the model arises from the use of distributed method of analysis, which account for areal distribution of many relevant factors. Its weaknesses are (1) its need for a large computer to simulate large watersheds, (2) the complexity of the data file necessary to describe the watershed, and (3) the assumptions such as zero sediment transport by subsurface flow, negligible channel erosion, and similar energy requirement of original and previously redetached sediment made for the simplified version of ANSWERS.

A modified version of the ANSWERS model fitted with the multilayer, transient conductivity Green and Ampt infiltration equation of Moore and Larson (1980) was used to predict runoff hydrographs from three small catchments of 0.07 to 3 ha in Toowoomba, Australia (Connolly et al., 1988). The original ANSWERS uses the Holtan infiltration model, which cannot simulate the sharp drop in crust infiltration that, if formed, results in increased runoff volume and peak flow. The Holtan model was replaced by a modified version of the Green and Ampt infiltration equation based on models presented by Moore and Larson (1980) and Brakensiek and Rawls (1983). This version of the Green and Ampt model is capable of representing the development of a surface seal during rain and the effect of differing soil layers (Connolly et al., 1988). Infiltration parameters were estimated using rainfall simulation data, direct measurement, and tabulated values. ANSWERS accurately predicted runoff hydrographs for all three watersheds. No data fitting or calibration was needed. Runoff prediction by ANSWERS was most sensitive to soil moisture deficit, surface soil conductivity, fraction of the soil surface exposed, and Manning's n.

2.3.12 Colorado State University Models

Li (1974) and Simons et al (1975) developed a high-resolution watershed sediment and routing model that is primarily applicable for surface erosion simulation. It is a distributed model that uses a grid and channel system to represent the watershed. This model simulates the land surface hydrologic cycle, sediment production, and water and sediment movement on a small watershed. It steps through any storm in small time increments and is generally used to analyze the response to only a limited number of storms. Conceptually the watershed is divided into an overland flow part and a channel system part. In the overland flow part processes of interception, evaporation, infiltration, raindrop impact detachment of soil, erosion by overland flow, and overland flow water and sediment routing to the nearest channel are simulated. In the channel system part water and sediment channel erosion or sediment deposition through the channel system is determined. A flow chart presenting the interrelationship of these processes is shown in Figure 2.10.

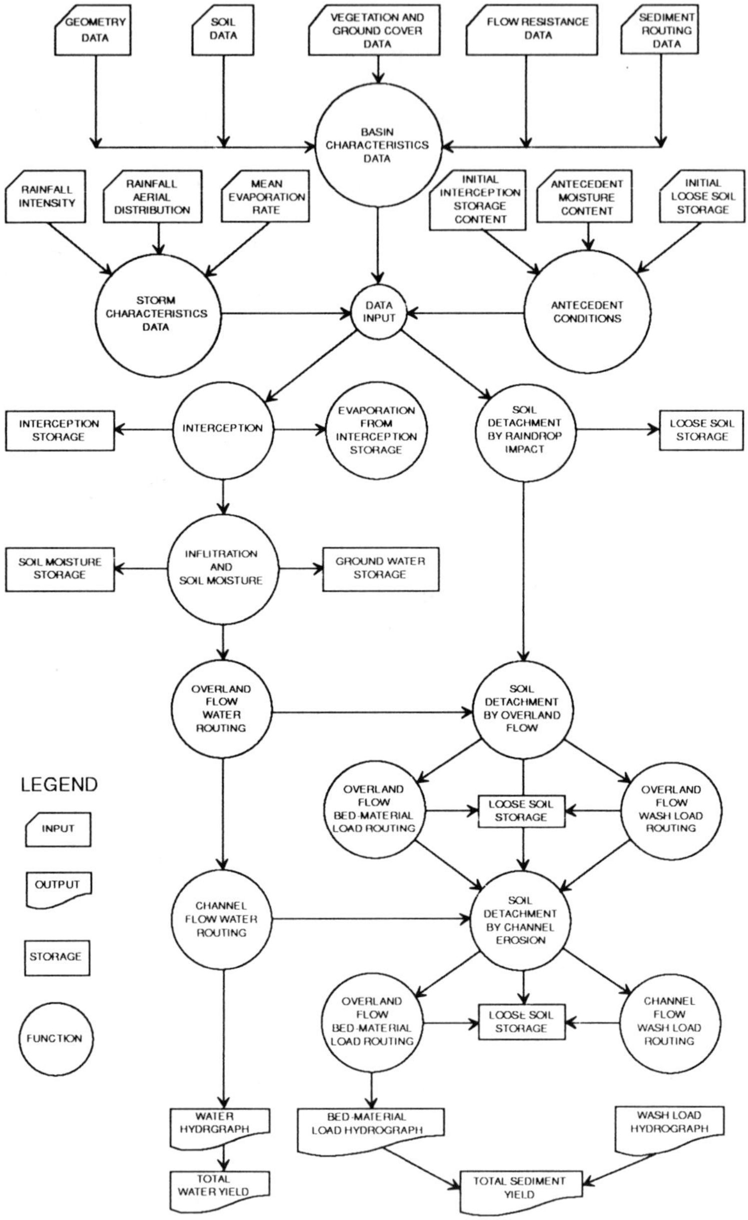

FIGURE 2.10. Flow chart for the watershed sediment and routing model. [From Simons and Li, 1980. In: Environmental Impact of Nonpoint Source Pollution. M. R. Overcash and J. M. Davidson (Eds.). Ann Arbor, MI: Ann Arbor Science, pp. 341–374. With permission.]

The channel component of the model includes an approximation to the dynamics terms of the full momentum equation. Even so the model cannot analyze a downstream flow control. Although values are available for most parameters, some calibration is required. because it leans heavily on concepts from noncohesive sediment transport, some of its relationships do not apply directly to agricultural field situations (Foster, 1980). Detachment and transportation of aggregates are not considered. Because the effect of cover and management on detachment by raindrop impact is related to canopy and ground cover but not to the soil characteristics the model is not suitable for the evaluation of the effect of crop rotations on soil erosion. Some erosion relationships have been incorporated in this model by Khaleel et al. (1979a,b) that better reflect detachment and transport on agricultural overland flow areas.

A simplified version of this model combines grid elements on uniform areas to reduce computations and to remove some of the limitations of a grid model (Simons et al., 1977; Simons and Li, 1980). The simplified model requires watershed to be represented by a channel and two contributing planes or a combination of two- and single-plane watersheds connected by a channel system. This much simpler representation of the watershed geometry provides for easier application but it is only applicable to homogeneous watersheds and may cause problems if the watershed is nonhomogeneous.

2.3.13 PWM

The Pollutant Washoff Model (PWM) is a physically based mathematical model proposed by Akan (1987) to predict pollutant washoff from impervious urban areas. Urban runoff models are generally simpler than those developed for agricultural areas because little or no consideration is given to infiltration and there is no need to update soil moisture content, crop growth, and related factors. Akan's model is based on the kinematic overland flow and convective pollutant transport equations. Detachment rate at any point is assumed to be proportional to overland flow bottom shear at that point with the proportionality constant depending only on the type of pollutant.

The model is based on three equations, one each for flow over impervious surface, pollutant washoff, and pollutant transport by overland flow. These equations are written in dimensionless form and are solved using an implicit finite difference scheme.

Because it addresses only a single element of the complex process of stormwater pollution, it can only be used as part of a more comprehensive urban runoff model (such as the Stormwater Management Model (SWMM) or Pesticide Runoff Model (PRM) described in Sections 2.5.14 and 2.5.16, respectively).

A strength of Akan's model is its ability to account for different rainfall intensity distribution in single-storm simulation. The model is verified only by a limited set of experimental data, so should be treated cautiously.

2.3.14 The Curtis Model

The Curtis model (Curtis, 1976) is a deterministic model for urban storm water and sediment discharge. It is a distributed model that time steps through the storm and uses the Meyer and Wischmeier's detachment and transport algorithms (Meyer and

Wischmeier, 1969). It uses overland flow planes and channels to represent the watershed. Selection of the elements to represent a given watershed, especially one with concave slopes, involves considerable judgment and experience. Well-defined guidelines for selection of elements are difficult to develop and error of judgment makes the results obtained by different users analyzing the same area significantly different. It is oriented primarily to urban watersheds and its major limitations are the need for calibration, and the fact that channel erosion is not included.

2.3.15 WEST

The Watershed Erosion and Sediment Transport (WEST) model is deterministic and operates in two phases (Leytham and Johanson, 1979). In the first, hydrologic and erosion processes occurring on land surfaces are simulated using the Agricultural Runoff Management (ARM) model (described in Section 2.5.3). The second simulates the movement of water and sediment through channels. The channels are represented by a network of ridges and the hydraulic routing is performed using the kinematic wave approximation. The results are passed to a sediment transport algorithm.

The model is designed to handle any mixture of cohesive and cohesionless sediment. It was developed by linking the ARM model to a channel routing model called CHANL. The uniqueness of WEST among the models developed up to that time is that it combines continuous simulation and channel routing, previously not done because of excessive computation costs. ARM (described in section 2.5.3) and its channel routing model, CHANL, run separately. CHANL uses an ARM output file as input. Data handling is difficult and considerable testing is needed to determine the usefulness and reliability of WEST.

2.3.16 Model for Lands Receiving Animal Waste

A mathematical model for sediment transport from lands receiving animal waste has been proposed by Reddy et al. (1979a,b) and Khaleel et al. (1979a,b). It uses generally accepted erosion and sediment transport modeling concepts. It is capable of simulating time-dependent erosion rates for various particle types during a storm event, erosion rate along the slope, and amount of sediment and manure coming from, on the surfaces, and within the soil surface layer. The model requires input from a deterministic hydrologic model capable of producing temporal and spatial estimates of flow discharge and depth of flow for a given storm, and the Colorado State University model (CSU) described by Li et al. (1977) has been used for this purpose.

The Khaleel model uses the Foster and Meyer (1972) concept of rill and interrill erosion. The Yalin sediment transport equation has been used for bed-load transport, and a modified version of Einstein's equation for suspended load (Li et al., 1977) in this study.

The model is based on the assumption that sediment load is the lesser of the transport capability of flow or the availability of material for transport. It simulates erosion over time and space during a storm, recognizing the two application modes of surface application and incorporation, for land areas receiving wastes. It divides erosion into rill and interrill detachment and it considers transport of four discrete particle types of fine primary, coarse primary, soil aggregates, and manure particles varying in degree of decomposition and ranging in size from colloidal humus particles to very large chunks. It also incorporates factors for soil, cropping management, ground cover, land use history, and tillage practices in the relationship for rill and interrill erosion.

For surface-applied manure the eroded solid in interrill areas is assumed to come from the soil and manure at rates proportional to their relative exposed areas. The total interrill solid loss is the sum of soil detachment rate and manure detachment rate. If no manure is applied all material detached originates from the soil and vice versa. A similar fractionation procedure has been used for rill areas. For incorporated manure a constant factor of 0.533 is used to modify the soil erodibility factor to a soil-manure detachability factor. This is based on Mazurak's finding that soil-incorporated manure increased soil detachment by raindrop impact by 0.533 mg m^{-3} h^{-1} for each metric ton of manure applied (Mazurak et al., 1975). However, the variable erodibility of different soils after manure application means that the factor is not constant. This puts severe limits on the applicability of the model to watersheds other than those for which it has been calibrated. Limitations may also arise from the lack of methods for estimating parameters such as the manure cover, manure erodibility, shear strength, particle size, and density of applied manure or soil-manure mixture.

2.3.17 Soil Erosion Model of Morgan et al.

Predictive models such as USLE or CREAMS need considerable input data, and are either very empirical (e.g., USLE), or complicated (CREAMS) (Morgan et al., 1984). The model of Morgan et al. (1984), which is more physically based than USLE and less complicated than CREAMS, uses an amalgam of existing relationships and is not a completely new model. It comprises a water phase and a sediment phase (Figure 2.11). In the sediment phase, which is a simplification of the soil loss equation described by Meyer and Wischmeier (1969), erosion is assumed to result from detachment of soil particles by rain splash and their transport by runoff. Splash is related to rainfall energy and rainfall interception; runoff transport capacity is related to the volume of runoff, slope steepness, and crop cover, as the following operating functions of the model show.

$$E = R_{\mathrm{a}}\left(11.9 + 8.7 \log I_{\mathrm{r}}\right) \tag{30}$$

$$Q_{\mathrm{u}} = R_{\mathrm{a}} \exp\left(-R_{\mathrm{c}} / R_{\mathrm{o}}\right) \tag{31}$$

$$S_{\mathrm{d}} - D_{\mathrm{s}}\left(Ee^{-aPr}\right)^{b} \times 10^{-3} \tag{32}$$

$$G = C_{\mathrm{c}} Q_{\mathrm{u}} (\sin\ \mathrm{S}) \times 10^{-3} \tag{33}$$

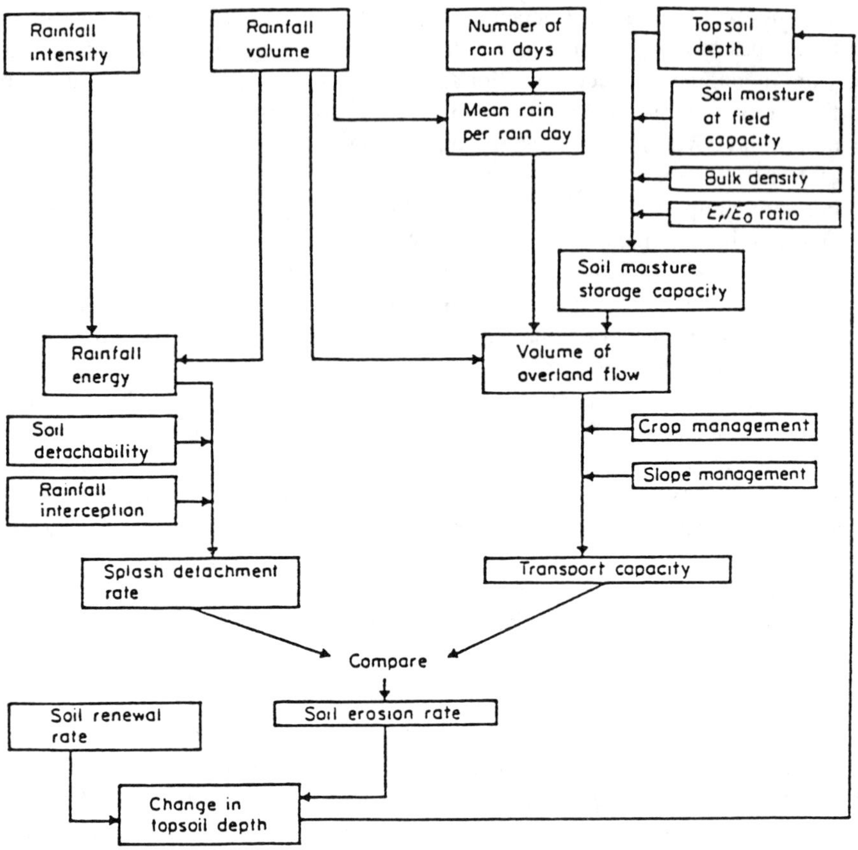

FIGURE 2.11. Flow chart of the model of Morgan et al. (From Morgan et al., 1984. *J. Agric. Eng. Res.* 30:245–253. With permission.)

where E is kinetic energy of rainfall (J m^{-2})

Q_u is volume of overland flow (mm)

S_d is rate of splash detachment (kg m^{-2})

G is transport capacity of overland flow (kg m^{-2})

R_a is annual rainfall (mm)

R_c is critical depth of daily rainfall to cause runoff (mm)

R_0 is R_a/R_n, where R_n is the number of rain days in the year

I_r is typical value for intensity of erosive rain (mm h^{-1})

D_s is soil detachability index (g J^{-1}), defined as the mass of soil detached from the soil mass per unit of rainfall energy

P_r is percentage rainfall contributing to permanent interception and stemflow

C_c is crop cover factor (a combination of C and P factors in the USLE)

S is slope angle

a and b are nondimensional constants

In the water phase annual rainfall is used to determine the energy of rainfall for splash detachment and runoff volume. The model compares the predicted rate of splash detachment with the transport capacity for overland flow and equates the rate of soil loss to the lower of these two values, thereby indicating whether detachment or transport is the limiting factor.

The results of a sensitivity test show that predictions are most sensitive to changes in annual rainfall and soil properties when erosion is transport limited, and in rainfall interception and annual rainfall when erosion is detachment limited. Validation of the model was carried out using published data for 67 sites in 12 countries where agricultural soil erosion has been measured on hillside erosion plots (Morgan and Finney, 1982). Prediction was not realistic for very low and very high rates of erosion. It successfully predicted runoff for 33 out of 56 test sites, and soil loss for 47 out of 67 test sites, giving respective success rates of 59 and 70%.

2.3.18 SPUR

The SPUR (Simulation of Production and Utilization of Rangelands) model (USDA, 1983) uses the SWRRB (Simulation for Water Resources in Rural Basins) model as a basis for its hydrologic component. It incorporates several improvements in routing water and sediment and facilitates spatial variability of precipitation, topography, and management practices (Lane, 1982; Wight, 1983). SPUR subdivides a watershed into fields, channels, and, if present, ponds. Fields can be defined by topography or land use. The only constraint on field is that it must be connected to a channel because the model cannot account for runoff-runon. A channel must either have an upland input (field) or two input channels. In this manner large watersheds can be simulated (Springer et al., 1984). Water and sediment are routed through channels. Ponds or small impoundments are allowed within the channel systems. Renard et al. (1983) present a more complete description of the conceptualization of the watershed by this model.

Overland runoff is estimated by a modified SCS Curve Number Equation (USDA, 1972) and upland soil loss is calculated using the Modified Universal Soil Loss Equation (MUSLE) of Williams (1975b). The routing of water and sediment through channel systems is done by procedures described by Lane (1982a,b).

Model testing carried out by Springer et al. (1984) revealed that the hydrology component of SPUR could be calibrated to predict both monthly runoff volumes and timing in an adequate manner. The prediction of sediment yield was rather poor. The sediment transport equations of SPUR have been derived assuming transport-limiting conditions while in the test case channels were sediment limiting.

2.3.19 SEDIMOT

SEDIMOT is an operational model developed at the University of Kentucky for evaluating the effects of land disturbances on the hydrologic regime, and to design sediment and runoff control facilities and structures (Wilson et al., 1986). It is a distributed parameter model with hydrologic and sedimentology components. The hydrology component consists of rainfall, runoff, and flow-routing modules. Rainfall excess is calculated by the SCS Curve Number method, and then routed to the watershed outlet using unit hydrograph techniques. Muskingum's routing method (Wilson et al., 1982) is used to route the resulting runoff hydrograph through a channel.

The accuracy of the hydrologic component was evaluated on several different watersheds and storms, and the agreement between the predicted and observed hydrograph was excellent to good. The sedimentology component utilized a modified version of the USLE (MUSLE), or detachment-transportation concepts incorporated into an algorithm called SLOSS. SLOSS is predicted by the relationship given in CREAMS.

To evaluate the effectiveness of different sediment control structure, the sediment graph and particle size distributions associated with sediment yield must be calculated. SEDIMOT-II is available in the form of either a batch or interactive mode of operation on most mainframe computers and some microcomputers. All model inputs are either available from standard hydrologic tables or are given in the user's manual (Wilson et al., 1982).

2.3.20 AGRUN

The Agricultural Runoff model (AGRUN) is a single-event simulation model derived from SWMM for nonurban application (Donigian et al., 1977). It takes precipitation data at 15-min intervals, and daily data for other parameters, as input. It is applicable to small and large multiple catchments. Watersheds can be divided into as many as 200 subareas of different sizes, and tributary drainage systems may have up to 200 channels. Overland flow across land surface and in small stream channels, groundwater interflow, and channelized tributary flow are taken into account. It uses the USLE for sediment erosion estimates. AGRUN is capable of simulating watersheds with different fields and crop types. This model simulates only sediment-attached residuals.

2.3.21 PRMS

The Precipitation-Runoff Modeling System (PRMS) was developed by Leavsley et al. (1983). It is similar to ANSWERS but more suitable for larger basins. The strong points about PRMS are in snowmelt and in representing channel hydraulic components for estimating flow peaks (DeCoursey, 1985).

2.4 EROSION-PRODUCTIVITY MODELS

2.4.1 PERFECT

The Queensland Department of Primary Industries (QDPI) has developed a computer simulation model [Productivity, Erosion, and Runoff Functions to Evaluate Conservation Techniques (PERFECT)] to be used in the subtropical region of Queensland and northern New South Wales (Littleboy et al., 1989). PERFECT is a hybrid model incorporating components from several existing models such as CREAMS (Knisel, 1980), EPIC (Williams, 1983), SPAW (Saxton et al., 1984), WBAL (Rosenthal et al., 1976), and SWRRB (Williams et al., 1985). It is a system model simulating plant-soil-water-management systems through time. It is a one-dimensional model representing a single hydrologic soil-crop unit and uses a time step of 1 day. It has been developed to simulate the major effects of management and environment and to predict runoff, soil loss, soil water, drainage, crop growth, and yield. An idealized representation of PERFECT is given in Figure 2.12.

The hydrology submodel of PERFECT is based on the Williams and LaSeur (1976) and Ritchie (1972) water balance models as used in CREAMS and similar models. PERFECT allows the user to select from a number of optional submodels to estimate soil erosion. These submodels include both versions of the modified universal soil loss equation (MUSLE) described before, a relationship developed by the QDPI and the simplified version of the GUESS model (Carroll et al., 1986).

PERFECT contains a rather large number of empirical components that must be validated with local data. Soil erosion is only a part of the system represented, and erosion component is a simple-event soil loss equation like MUSLE (Williams, 1975b). Since this model uses a daily time step, accurate prediction of runoff rates and rainfall energy is difficult as the time distribution of rainfall within each day is

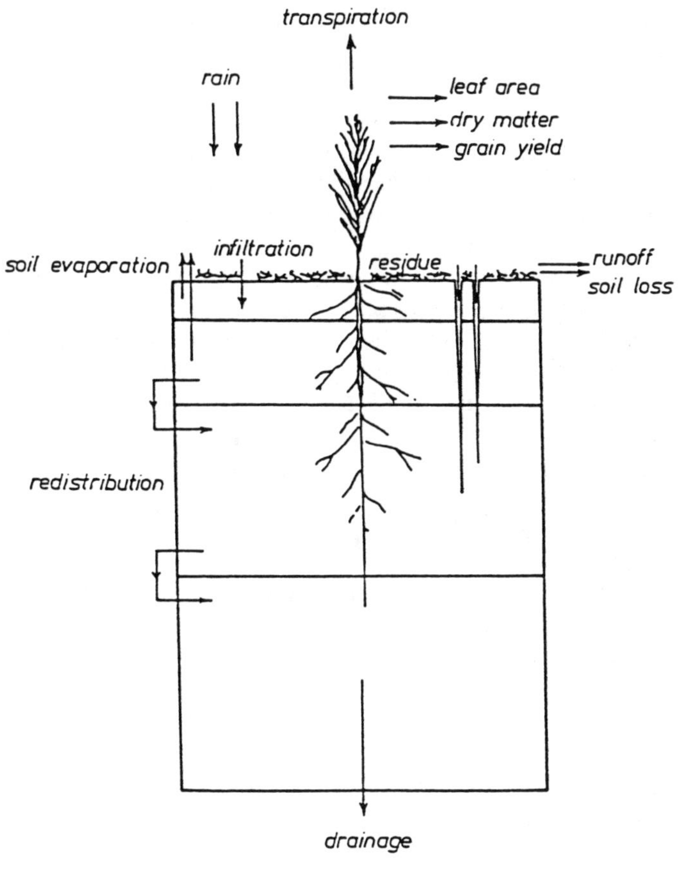

FIGURE 2.12. Idealized representation of PERFECT model. (From Littleboy et al., 1989. Bulletin QB89005. Brisbane, Australia: Queensland Department of Primary Industries. Queensland, Australia. With permission.)

hard to obtain. This limits the accuracy of prediction of event soil loss as variation in event soil loss is in part due to variations in runoff rate and rainfall energy. However, this model is an advance on similar models since it incorporates interactions between management strategies, hydrological processes, and a fully dynamic crop model. The same model can be used for short-term calibrations as well as long-term simulations. PERFECT is written in FORTRAN and the user can select different algorithms to estimate soil erosion, runoff, and drainage.

2.4.2 EPIC

The EPIC model (Erosion-Productivity Impact Calculator) was developed by Williams et al. (1983) to obtain estimates of the long-term effects of soil erosion on productivity (Williams, 1983). This mechanistic model is application oriented and provides a means of rapidly evaluating various management schemes, soil types, and climate sequences. The model can be divided into nine main areas: hydrology, weather, erosion, nutrients, plant growth, soil temperature, tillage, economics, and plant environment control.

Hydrology is predicted using the Williams-Ritchie model with minor changes to the runoff submodel. Soil erosion is estimated using a MUSLE based on runoff depth and peak rate, rainfall energy, and catchment conditions. Profile depth can be adjusted for depth of soil eroded. The model also incorporates a comprehensive nutrient submodel. Crop growth is calculated from user-defined parameters that describe the leaf area development over time. Generalized yield prediction equations for a large range of crops are used to estimate crop yield. Weather data can be either inputed or generated by the model.

The detail in the model is sufficient enough to allow the determination of general trends in soil erosion and crop yields but, being a more empirical than process-driven model, its ability to evaluate comprehensively the relationships between eroded soil conditions and crop productivity is limited. Other erosion-productivity models such as GOSSYM (Baker et al., 1983) and NTRM (Shaffer and Larson, 1982; Shaffer and Pierce, 1982; Shaffer et al., 1983) are more process-based models than EPIC and have the capabilities at the research level to simulate complex interactions of growing crops with various soil properties.

2.4.3 SPAW

The SPAW (Soil-Plant-Air-Water) model was developed to predict soil water and crop productivity interactions (Saxon et al., 1984). The strength of SPAW lies in the soil water redistribution component. Daily time steps are used except in redistribution, which is calculated many times for each day. Daily infiltration is assumed to occur instantaneously and the Darcy equation is used to predict water movement through the profile. Runoff volume is predicted using either the curve number method or is input from an observed runoff data set. The crop growth and yield prediction submodels require numerous parameters.

2.4.4 SWRRB

The SWRRB (Simulation for Water Resources in Rural Basins) model was developed to investigate management decisions for ungauged rural catchments (Williams et al., 1985). Hydrology is driven by the Williams-Ritchie-type model using the CREAMS daily rainfall and hydrology model. The main enhancement of SWRRB is the division of a natural watershed into a number of smaller subcatchments representing different soils or topography. Hydrology calculations are performed simultaneously across all subcatchments with a simple routing algorithm to obtain catchment runoff. Crop growth and yield are simulated dynamically with general functions producing leaf area and biomass. The crop model accounts for water and temperature stress.

2.4.5 NTRM

The Nitrogen Tillage Residue Management (NTRM) model is a mathematical model developed to simulate the impact of soil erosion on short- and long-term productivity of soil (Shaffer, 1985). A generalized block diagram of the model is shown in Figure 2.13. The model is written in modular form with submodels for the various processes considered to be linked together to form the overall system. NTRM requires a large number of inputs, which include the following:

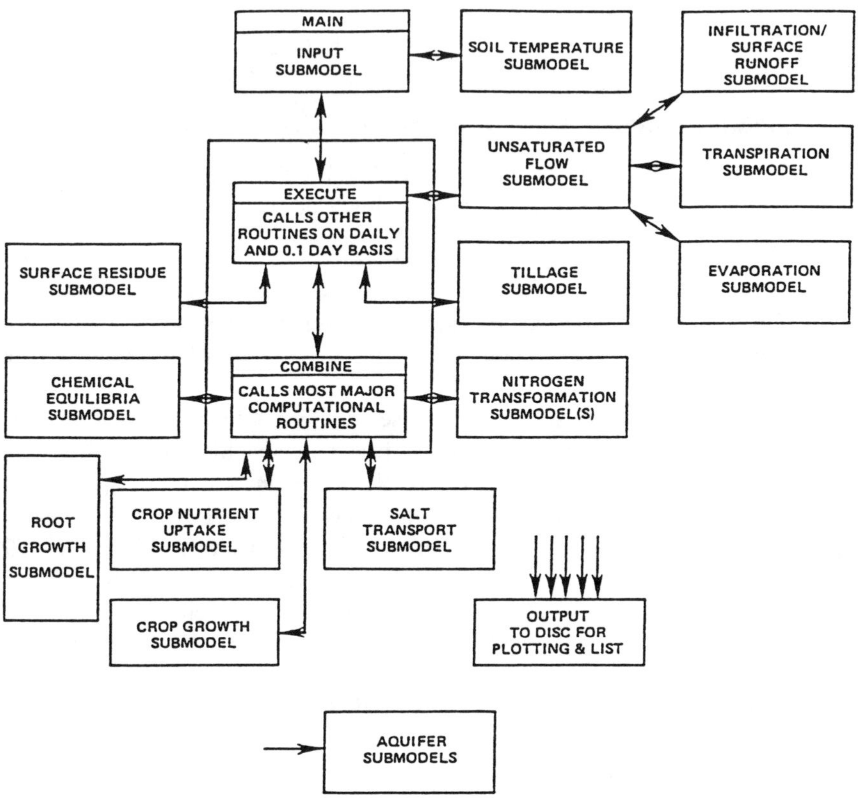

FIGURE 2.13. NTRM model diagram. (From Shaffer, 1985.)

1. Climate variables such as daily precipitation, maximum and minimum air temperature, pan evaporation, wind speed, and solar radiation
2. Soil properties such as soil-water characteristic curves, water content, bulk density, texture, roughness, nutrients and salt concentration, soil temperature, soil profile geometry, and slope length, percent, and aspect
3. Management inputs such as the date and type of tillage, irrigation, fertilizers, and crop residue, and crop population, emergence, and maturity date

Model validation for crop yield prediction, with the data from the literature, suggested that the long-term model prediction of crop yield on a variety of soils and environmental conditions was in a reasonably good agreement with the measured data. The same validation test showed that the short-term predicted yields differed significantly from the long-term ones, suggesting some limitations in using the model for the time periods shorter than 10 years (Shaffer, 1985). Because the NTRM model is a research-level tool, total annual simulations for a long period could be very costly or time consuming. In addition to predicting yields under noneroded and eroded conditions, the NTRM model can be used to assess the relative benefits of various management practices.

2.5 MODELING SORBED CHEMICAL TRANSPORT

2.5.1 Loading Functions for Nutrients and Pesticides

If a pollutant is assumed to be linearly correlated with the loss of sediment from a watershed, the pollutant load (L_n) can be represented by

$$L_n = P_n \cdot S_1 \tag{34}$$

where S_1 is load of sediment delivered, and P_n is a potency factor of pollutant n (mass of pollutant/mass of sediment) (Dean, 1983). This expression is known as a loading function.

The loading function approach is easy to use and does not require extensive data. The value of S_1 in the above equation is usually based on the USLE and the SCS curve number for runoff. The USLE and the Curve Number Equation are not designed to evaluate nonpoint source pollution although they have been used extensively for this purpose, as reported by Haith and Dougherty (1976) and Haith and Tubbs (1979). The ability of loading functions based on such estimates to predict nonpoint source pollution in large agricultural watersheds has, therefore, not been established and a consistent methodology for applying loading functions to a watershed is not available (Haith and Tubbs, 1981).

The potency factor (P_n), which is the essence of the loading function, is a complex factor combining the effects of many processes and is difficult to estimate accurately. An approximate expression used by Dean (1983) for P_n is

$$P_n = \mathrm{ER}_n \cdot C_a \cdot f \tag{35}$$

where C_a is the average concentration of pollutant n in the surface layer of the soil (i.e., mass of pollutant/mass of surface soil), ER_n is the enrichment ratio of pollutant n (mass of pollutant in eroded sediment/mass of pollutant in surface soil), and f is the ratio of the mean particle densities of surface soil to that of eroded sediment. The enrichment ratio (ER) may well be involved in sediment delivery but conceptually should not be associated with a "potency factor," as is commonly the case (Dean, 1983). The enrichment ratio represents the effects of several processes including preferential detachment and transport of fine particles such as organic matter, clay, and silts. For strongly adsorbed pollutants the enrichment ratio of fine soil particles in runoff will have an impact on their loss, and for less strongly adsorbed ones this effect diminishes (Dean, 1983). The values computed for ER by Baker and Johnson (1978), Leonard et al. (1979), and Baker (1980) suggest a relationship similar to the one discussed above between enrichment factor and the degree of adsorption of a compound. Potency factors or enrichment ratios are being used in many models including NPS, HSPF, CREAMS, AGRUN, ACTMO, and WASCH (each will be discussed in this chapter).

Haith and Tubbs (1981) integrated available simple methods into general loading functions to estimate mass fluxes of dissolved or solid-phase pollutants in watershed runoff. They tested their loading functions by comparing predicted and observed nitrogen and phosphorus mass flux in stream flow from a large watershed in Pennsylvania for a 16-month period. The comparison indicates that loading functions provide a reasonable means of estimating agricultural nonpoint sources but there are some limitations. Snowmelt runoff was not modeled accurately and there is uncertainty in selecting appropriate delivery ratio for solid-phase pollutant.

Menzel (1980) developed regression equations for enrichment ratio on sediment discharge for nitrogen and phosphorus under various crop management, and Zison (1980) developed potency factors for a wide range of pesticides, nutrients, and conventional pollutants. Potency factors and ER are most widely researched and documented for the nutrients nitrogen and phosphorus in agricultural lands. Some empirical equations are now available for estimating ER_N and ER_P, the enrichment ratios for nitrogen and phosphorus.

ER_P values required for modeling phosphorus in runoff can be obtained by measurement or estimated from organic matter and clay enrichment ratios (Logan and Stiefel, 1979). ER_P is higher for low rates of erosion and for coarser-textured soils (Porter, 1975; Logan and Stiefel, 1979). According to Stoltenberg and White (1953) ER_P varies between 1 and about 3.5.

ER_P values have also been used with the USLE values for gross erosion and sediment delivery values to model the effect of catchment erosion on the phosphorus load of streams and basins (McElroy et al., 1976). Such a model, however, does not adjust ER_P with the gross erosion changes and therefore it would overestimate the effect of reduced gross erosion on phosphorus loads.

The enrichment ratios for organic matter (ER_{OM}) range from one to five, higher in sandy soils and lower in fine textured, highly erodible soils. Frequently it is assumed that ER_{OM} is similar to the clay enrichment ratio. Measured data on soil organic matter content are usually available, but some workers use 20 times total nitrogen when local data are not available.

Potency factors and ERs for pesticides have not received much attention. Leonard et al. (1979) developed the following equation for predicting the total concentration of four soluble and cationic herbicides in runoff:

$$C_p = a(C_{sp})^b \tag{36}$$

where C_p is the total pesticide concentration in runoff (ppm), C_{sp} is the concentration of pesticide in 0 to 1 cm of top soil (ppm), and a and b are regression coefficients. For four soluble pesticides studied, b was found to be equal to 1.20, being 0.83 for more strongly adsorbed paraquat. Coefficient a, identified as an enrichment factor, had a value of 2.16 for strongly adsorbed and 0.05 for weakly adsorbed herbicides. In a similar study Baker (1980) found $a = 0.095$ and $b = 1.03$, values very close to those of Leonard et al. (1979). Dividing both sides of Leonard's equation by C_{sp} gives the enrichment ratio (ER), which is then shown to depend on the pesticide concentration in the soil.

Very little work has been done to evaluate enrichment ratios of pesticides, and then only for a few compounds. There is a need to discover how ER relates to more well-known physical or chemical properties and how it is affected by erosion processes.

Use of the potency factor concept is not appropriate for substances that do not travel physically bound to sediment. In these cases modeling pollutant losses as a direct function of runoff or leaching by infiltration is more realistic.

In order to use the loading function concept the sediment load must be known. The USLE is the method that has been most commonly used for this purpose. CREAMS uses a modified version of USLE, while in NPS and HSPF (Table 2.1) power functions of rainfall intensity and flow are used to simulate the detachment and transport processes. Potency factors must be supplied by the user by measurement or extrapolation from published data. Based on the degree of accuracy required, the enrichment ratio can be assumed constant (for strongly adsorbed substances being approximated by the soil fines enrichment ratio), or be evaluated as a function of sediment production with ER decreasing as sediment yield increases.

2.5.1.1 Transport of Nitrogen

Losses of nitrogen from agricultural land occur by several modes of transport and for several forms of nitrogen. Loss by leaching can be substantial (Winteringham, 1977), especially for inorganic forms of nitrogen, losses being chiefly as nitrate (White, 1988). Leaching losses can be modeled (Frere, 1975; Rose et al., 1982a,b; White, 1988), but this is a topic not covered by this book. Losses of nitrogen can also be in gaseous form following volatilization or denitrification. Since the recent development of simplified micrometeorological methods of measuring gaseous losses of nitrogen their significance has been considerably clarified (Freney and Black, 1988).

Losses in runoff of surface-applied nitrogen fertilizer can be significant (Haith and Dougherty, 1976; Rose and Dalal, 1988), but models of such runoff losses are empirically plausible (Young et al., 1987) without being well based, and the effect of management factors is not well understood (Dean, 1983).

With the exception of nitrogen in nitrate form, or converted to gaseous forms, the loss of nitrogen is chiefly associated with the loss of sediment (Frere et al., 1975). Thus the rate and mechanism of loss of nitrogen is strongly affected by the transformations between the various organic and inorganic forms of nitrogen that are continuously taking place. Thus the vast literature on nitrogen transformations and nitrogen cycling (recently reviewed in Wilson, 1988) is a necessary background to modeling losses of nitrogen in the various processes briefly referred to above.

2.5.1.2 Transport of Phosphorus

Models used for predicting phosphorus in runoff must account for the transport of particulate and soluble phosphorus and for exchange (adsorption/desorption) between them (Logan, 1980). One approach is to analyze the surface soil for particulated and soluble phosphorus, route the soil water and its soluble phosphorus load to the stream, and separately route the eroded sediment to the stream. A phosphorus enrichment ratio (ER_P) must be used to relate phosphorus concentration in the original soil to the phosphorus concentration in eroded sediment. Reported measurements of ER_P vary from 1.00 to about 3.5 (Stoltenberg and White, 1953; Syers et al., 1969), and can be predicted from organic matter and clay enrichment ratios (Logan and Stiefel, 1979). ER_P value is greater at low rates of erosion and for coarser-textured soils (Porter, 1975; Logan and Stiefel, 1979). This approach assumes zero exchange between sediment and soluble phosphorus, an assumption that is not completely valid for soil-water systems. Dilution of pore water with rain water of lower phosphorus concentration can result in phosphorus desorption from the sediment. Chemical transformation in the aquatic environment and the exchange between soluble and sorbed nutrients have been reported by many workers (Frink, 1969; Patrick and Mahapatra, 1968; Syers et al., 1973).

Another approach for modeling phosphorus in runoff is to incorporate a desorption isotherm to predict change in the soluble phosphorus load during runoff (Logan, 1980). It can be assumed that the original equilibrium concentration (OEC) will be attained by the time runoff enters the stream, and so OEC times the water volume will be the soluble phosphorus load. This approach assumes that

1. There is sufficient labile sediment phosphorus for equilibrium to be reached during runoff.
2. The rate of desorption is fast enough for the equilibrium to occur during runoff.

If sediment concentration is low, or runoff duration is short, and therefore the above assumptions do not apply, it might be possible to estimate labile phosphorus content of sediment and to assume that it is all desorbed during runoff and to use a first-order decay equation, such as

$$A_P = A_u e^{-kt} \tag{37}$$

where A_u is labile phosphorus pool prior to runoff, k is decay rate, and t is runoff period.

A phosphorus loss-estimating submodel has been developed by Storm et al. (1988) to be incorporated into the ANSWERS watershed model to account for soluble and sediment-bound phosphorus transport. The extended version of ANSWERS, which is capable of simulating the transport of individual particle size classes, was used as their sediment transport model. The phosphorus transport submodel has the three components of soluble phosphorus (P_{sol}) desorbed from the soil surface into runoff, sediment-bound phosphorus (P_{sb}), and the equilibrium between the soluble and sediment-bound phases.

Sediment-bound phosphorus is modeled as a function of the specific area of the soil and transported sediment. Transport of P_{sb} is calculated using the following equation:

$$P_P = \sum_{i=1}^{n} P_i \times SED_i \tag{38}$$

when P_P is the total amount of P_{sb} transported, P_i is the phosphorus content of particle class i, SED_i is the eroded sediment transported by the surface runoff for particle class i, and n is the number of particle size classes. P_i is estimated by assuming that P_{sb} is distributed in the sediment in proportion to the specific surface area (SSA) of the particles, which can be determined experimentally or estimated from soil texture.

For estimation of soluble phosphorus concentration the desorption equation of Sharpley et al. (1981a,b) was modified to account for differences in phosphorus extraction technique by adding an empirical constant to the initial amount of desorbable phosphorus in the soil. The equation is as follows:

$$P_d = B \times t^{\lambda} \times WS^{\beta} \times P_a^{\gamma} \tag{39}$$

in which P_d is the cumulative phosphorus desorbed (mg kg^{-1}), t is contact time (min), WS is the water-to-soil ratio, P_a is available phosphorus content of soil (mg kg^{-1}), and B, λ, β, and γ are empirical constants for a given soil. These constants are dependent on soil characteristics and can be estimated experimentally or from an empirical equation developed by Sharpley (1983).

Model verification was carried out in the field on tilled and nontilled soils with rainfall simulator and the model predictions were compared with observed runoff, sediment, soluble phosphorus, and P_{sb} yields. Because ANSWERS was calibrated for the plots, simulated sediment yields were in good agreement with predicted ones. Sediment-bound phosphorus yield predictions were reasonably good but the agreement between measured and predicted soluble phosphorus was poor. A substantially higher observed ER_P value than that simulated is attributed to a low initial phosphorus estimate. Application of this model requires the estimation of soluble phosphorus and Langmuir isotherm coefficients for each watershed, involving a large amount of field data collection. The model (Storm et al., 1988) has also been used as a flexible planning tool for evaluating the cost effectiveness of various conservation tillage practices on watershed scale with some degree of success.

2.5.1.3 Transport of Pesticides

Among the mathematical models used to describe equilibrium pesticide adsorption-desorption in a soil water system, Freundlich and Langmuir equations have been most successful (Rao and Davidson, 1980). The Freundlich equation is of the form

$$S_a = K^* C_s^N \tag{40}$$

Where S_a is amount of pesticide adsorbed to the soil (μg g^{-1}), C_s is solution phase pesticide concentration (μg ml^{-1}), and K^* and N are empirical coefficients that depend on the soil and pesticide involved.

The Langmuir equation is of the form

$$\frac{C_s}{S_a} = \frac{1}{K_L S_{max}} + \frac{1}{C_s S_{max}} \tag{41}$$

where S_{max} is the maximum adsorption capacity of soil (μg g^{-1}), K_L is the Langmuir coefficient related to adsorption bonding energy (ml μg^{-1}), and S_a and C_s are as defined above. The Langmuir equation generally fails to describe pesticide adsorption in a complex and heterogeneous medium such as soil because it assumes that:

1. The energy of adsorption is the same for all sites.
2. Adsorption is localized with no interaction between adsorbed molecules.
3. The maximum adsorption capacity used in the equation is represented by a monolayer coverage.

The ability of the Freundlich equation to describe pesticide adsorption on soils at both high and low concentrations has been shown by several workers (Hamaker and Thompson, 1972; Davidson et al., 1978a,b; Rao and Davidson, 1979). Although the Freundlich equation has no conceptual basis, it can be derived from the Langmuir equation if the energy of adsorption is assumed to decrease with increasing surface coverage (Thomas and Thomas, 1967).

For solution concentrations similar to agricultural ecosystems, Langmuir and Freundlich equations often can be assumed to be linear with the following shape:

$$S_a = K_D C_s \tag{42}$$

with $N = 1$ in the Freundlich equation and $K_L C_s < 1$ in the Langmuir equation. K_D is commonly referred to as the adsorption partition coefficient.

The error introduced by assuming the adsorption isotherm to be linear may be tolerated for practical application, as in the NPS model (described in Section 2.5.2) for $0.1 < C_s < 10$ and $0.5 < N < 1.0$. However, for high solution concentrations, such as those encountered under pesticide waste disposal sites, it may cause a large overestimation of the amount of pesticide adsorbed (Davidson et al., 1978a,b; Rao and Davidson, 1979).

Most of the literature deals with the partitioning of pesticides that had achieved equilibrium between soil and solution phases, while a few have investigated the rate at which this equilibrium is achieved. Under steady or transient conditions of water flow in soil, sufficient time may not be available for equilibrium of pesticide adsorption-desorption to be achieved. Consequently many investigators such as Van Genuchten et al. (1974) and Haque et al. (1968), considered the kinetics of the adsorption-desorption processes, and have developed models on that basis.

The following expression was proposed by Van Genuchten et al. (1974) for the amount of pesticide absorbed with time, with the assumption that it follows reversible nonlinear kinetics:

$$\frac{dS_a}{dt} = K_2\left[\left(\frac{K_1\theta}{K_2\rho}\right)C_s^N - S_a\right] \tag{43}$$

where K_1 and K_2 are, respectively, the kinetic rate coefficients for forward and backward reactions, θ is soil water content, ρ is soil bulk density, t is time, and C_s and S_a are as defined earlier. At equilibrium $dS_a/dt = 0$ and this equation reduces to the Freundlich equation with $K^* = \theta K_1/K_2$.

Other adsorption-desorption kinetic models have been developed by Hague et al. (1968) and by Lindstrom et al. (1970). The former, working with the pesticide 2,4-D and clay minerals, assumed negligible reverse reaction (desorption rate) while the Lindstrom model allows for adsorption-desorption energy to change with surface coverage. These models are valid for homogeneous surfaces because a single adsorption-desorption mechanism is assumed to be responsible for adsorption. In soil there is a variety of reactive compounds, such as organic matter as well as clay minerals, with different adsorption-desorption energies and kinetics, and therefore this assumption may not be valid. Furthermore, adsorption sites may not be equally accessible. Diffusion of pesticide to adsorption sites in the interior of organic matter may take much longer than the rapid reaction on the surfaces of clay minerals.

For each runoff event a certain amount of applied pesticide is lost in soluble form as well as with the sediment in the runoff. The following equations have been developed by Mulkey and Falco (1977) for the estimation of the relative amounts of pesticide lost in each of the two phases:

$$F_w = \left(1 + \frac{c\overline{K}_D}{\rho}\right)^{-1} \tag{44}$$

$$F_s = \left(1 + \frac{\rho}{c\overline{K}_D}\right)^{-1} \tag{45}$$

and where F_w and F_s are the fractions of pesticide lost in the water phase and in association with the sediment phase, respectively. ρ is the density of water, c is the sediment concentration in the runoff (g ml^{-1}), and K_D is the average adsorption partition coefficient. In deriving these equations it has been assumed that adsorption-desorption isotherms are linear and that sorption equilibrium is achieved instantaneously. These assumptions put severe limitation on the application of these equations. Data collected by Rao and Davidson (1980), Leonard et al. (1979), and Smith et al. (1978) suggest that the pesticides are lost in the water phase of the runoff except for highly adsorbed pesticides ($K_D > 100$ g ml^{-1}) or for high sediment concentrations ($c > 1000$ g ml^{-1}). Total pesticide loss in each runoff event decreases exponentially with time, and is strongly correlated with the total amount of pesticide remaining in the runoff active zone (0- to1-cm depth) of the soil surface (Leonard et al., 1979).

2.5.2 NPS

The Nonpoint Pollutant Source (NPS) model is a mathematical model that continuously simulates pollutant contributions to stream channels from nonpoint sources. It is designed for application by planning agencies in analysis of nonpoint pollution problems (Donigian and Crawford, 1976a, 1977). It consists of three major subprograms, which represent hydrologic response of the watershed, and the processes of pollutant accumulation, generation, and wash-off from the land surface (Figure 2.14). The hydrologic algorithms have been derived from the Stanford Watershed Model (Crawford and Linsley, 1966) and the Hydrocomp Simulation Program (Hydrocomp, 1969).

This model can simulate nonpoint source pollution from a maximum of five different land use categories in a single operation. Pollutant parameters are specified separately for pervious and impervious areas within each land use and can vary with the month of the year to represent seasonal pollution problems. The NPS model algorithm is based on the concept that sediment and sediment-like material can be used as an indicator of the land surface contributions of many nonpoint pollutants, and these pollutants are related to sediment by a potency factor. It is therefore not suitable for the simulation of soluble pollutants. This model, like ARM (Section 2.5.3), does not simulate channel processes, and is applicable only to watersheds smaller than some 500 ha (2 mi^2) in which channel flow effects are assumed to be small. Due to lack of a complete quantitative description of the processes controlling nonpoint pollution, this model needs to be calibrated. Such model calibration has been discussed by Donigian and Crawford (1977). This model was later incorporated in the ARM model. It is also limited to the evaluation of land surface contributions to nonpoint source pollution, with subsurface and ground-water pollution not considered.

It has been assumed for both ARM and NPS models that sediment production by erosion is at the transport capacity over the long term, so that there is no net deposition. This is an unrealistic assumption, especially for concave slopes. Another disadvantage of these two models is that because they are lumped, spatial variability cannot be considered.

Model testing on three urban watersheds consisting of residential, commercial, industrial, and agricultural lands indicated good agreement between recorded and

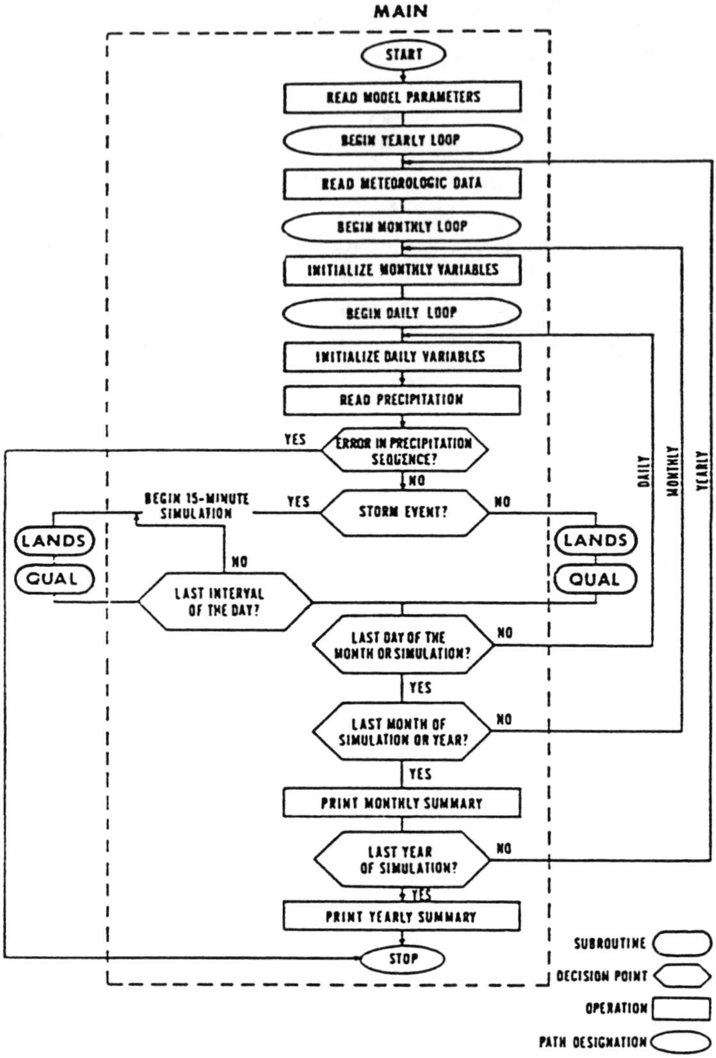

FIGURE 2.14. NPS model structure and operation. (From Donigian and Crawford, 1976b. Report No. EPA-600/2-76-043. U.S. Environmental Protection Agency. Athens, GA. With permission.)

simulated hydrology and pollutant wash-off, but there are many limitations in model calibration for watersheds as pointed out by Cahill et al. (1979). Only some of these limitations have been removed in the ARM model.

2.5.3 ARM

The most widely tested combined model for agricultural lands is EPA's ARM (Agricultural Runoff Management) model (Donigian and Crawford, 1976b,c). It is an independent version of the NPS model (Donigian and Crawford, 1976a). The ARM

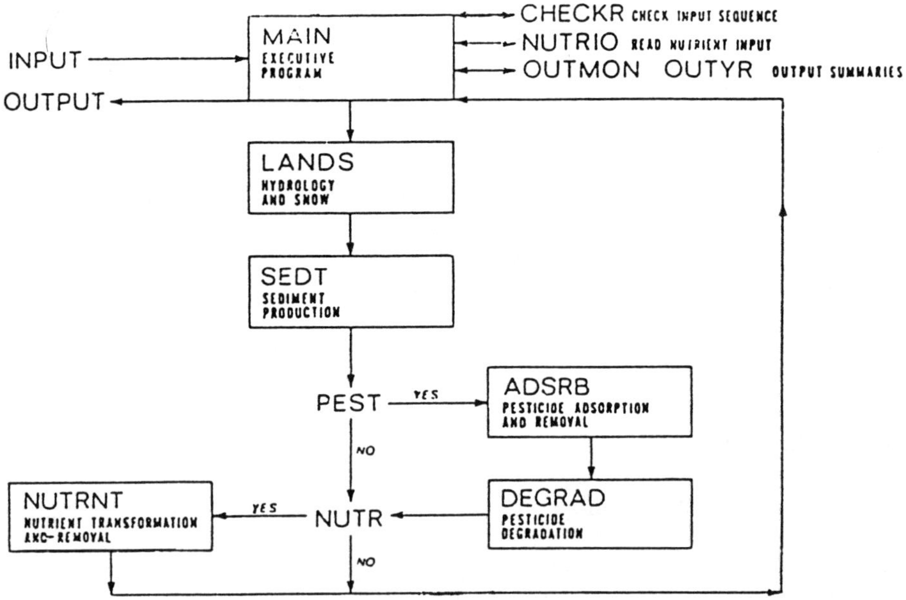

FIGURE 2.15. ARM model structure and operation. (From Donigian and Crawford, 1976b. Report No. EPA-600/2-76-043. U.S. Environmental Protection Agency. Athens, GA. With permission.)

model simulates runoff (including snowmelt), sediment, pesticides, and nutrient contributions to stream channels from both surface and subsurface sources. No channel routing procedures are included, so the model is applicable to watersheds small enough for channel processes and transformations to be considered negligible. Although the limiting area varies with climatic and topographic factors, the upper area limit of ARM applicability is about 1 or 2 mi^2 (250 to 500 ha) because channel process significantly affect water quality from larger (and perhaps smaller) watersheds.

As shown in Figure 2.15, the major components of the model are a hydrologic submodel (LANDS), sediment production submodel (SEDT), pesticide adsorption and removal (ADSRB), pesticide degradation (DEGRAD), and nutrient transformation submodel (NUTRNT). Total eroded soil mass is a necessary ingredient in the pesticide adsorption/desorption and nutrient transformation subroutines, and the depth of surface soil (or depth of chemical incorporation) requires specification in the model input parameters.

Results of tests on watersheds shows fair agreement between simulated and measured data for some pesticides (Donigian and Crawford, 1976b). For these tests, the groundwater zone has been considered a sink for deep percolating chemicals, and the groundwater flow contribution has been neglected. However, on larger watersheds, this contribution could be significant.

The mathematical foundation of LANDS was originally derived for the Stanford Watershed Model (Crawford and Linsley, 1966). Major parameters of the LANDS subprogram are identical to these of the Hydrocomp Simulation Program (HSP)

(Hydrocomp, 1969). Through a set of mathematical functions, LANDS simulates the major components of the hydrologic cycle including interception, surface runoff, infiltration, and percolation to groundwater. Soil loss algorithms are based on the work of Negev (1967) at Stanford University, but they were later influenced by the work of Meyer and Wischmeier (1969) and Fleming and Fahmy (1973). Only sheet and rill erosions are included in the sediment loss submodel because gully erosion was assumed insignificant in small watersheds. The two component processes of sheet and rill erosion pertain to (1) detachment of soil fines (generally silt and clay fractions) by raindrop impact and (2) pickup and transport of soil fines by overland flow.

The algorithm employed to simulate the process of pesticide adsorption/desorption is composed of an empirical term that accounts for pesticides permanently adsorbed to soil particles that will not desorb under repeated washing plus the standard Freundlich single-valued adsorption/desorption isotherm. The concentration of permanently adsorbed pesticides must exceed the adsorption capacity of soil for adsorption/desorption equilibrium to become operative. Others have shown that the assumption of single-valued adsorption/desorption is not valid for many pesticides (Davidson and McDougal, 1973; Van Genuchten et al., 1974; Davidson et al., 1973), and the algorithm used in the ARM model does not adequately represent the division of pesticides between sediment and solution phases.

General pesticide degradation and volatilization models for chemical, microbial, and photochemical degradation have been developed, but volatilization models based on the work of Farmer and Letney (1974) were not included in the last version of the ARM model due to lack of field data for testing purposes.

Nutrient simulation in the ARM model attempts to predict nutrient losses from erosion, surface runoff, leaching, and biological conversions. Major pathways in nitrogen and phosphorus cycles in the soil have been considered, and with extensive testing and calibration the model may be used to develop fertilizer management plans.

The ARM model, like the more complex HSPF model that incorporates it, requires extensive calibration. It seems possible to calibrate the model in such a way that it simulates nutrient losses accurately (Haith and Loehr, 1979).

2.5.4 ACTMO

The Agricultural Chemical Transport Model (ACTMO) was developed by the Agricultural Research Service of the USDA for chemical movement from complex watersheds (Frere et al., 1974, 1975). The initial version is designed for the small, farm-size watershed, which can be represented by a single rainfall input. It consists of three submodels for hydrology, erosion deposition, and chemical transport. The flow diagram for this model is given in Figure 2.16.

The hydrologic component is a modified version of the USDAHL-70 model of watershed hydrology described in Section 2.2.5 (Holtan and Lopez, 1971). The areal subunit or "zone" is constructed by grouping together fields of the same physical and crop management features. USDAHL and ACTMO are basically lumped parameter models. The model continuously accounts for soil moisture by balancing infiltration,

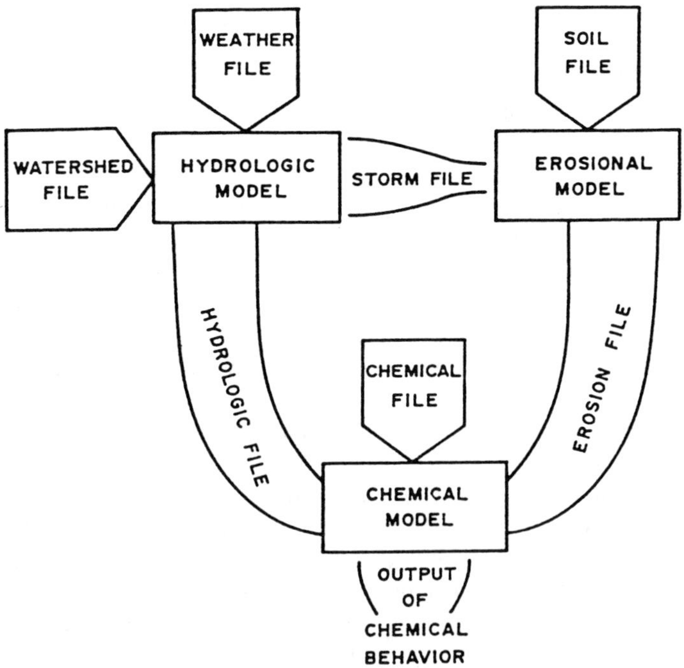

FIGURE 2.16. Flow diagram for the input-output files in the ACTMO model. (From Frere et al., 1975. ARS-H-3. Agricultural Research Service, U.S. Department of Agriculture, Washington, D.C. With permission.)

evapotranspiration, and seepage. Infiltration and rainfall excess are computed for each zone by comparing rainfall to infiltration capacity for each zone. Rainfall in excess of infiltration and surface storage is routed across each zone using a modified kinematic flow method.

The erosion-deposition and chemical submodels were developed specifically for this model, and combine many previously published concepts. The erosion-deposition submodel predicts soil loss by a modified version of the USLE that includes the effect of runoff on detachment and transportation of soil particles as proposed by Foster et al. (1977). The erosion submodel is limited to rill and interrill contribution and is capable of estimating particle size and distribution by calculating the clay enrichment ratio.

The chemical submodel traces the movement of a single application of a chemical through and over watersheds. The model simulates sorption/desorption processes. The sorbed pollutant fraction moves with the detached soil particles while surface runoff contains a dissolved phase. The initial version is limited to tracing a single application of one chemical at a time. The application of more than one chemical can be handled by repeating the chemical submodel for each chemical. This procedure assumes no interaction between chemicals, and no direct effect by chemicals on erosion or hydrologic behavior of watershed, which may not be the case because fertilizer and pesticides affect crop cover, which in turn affect hydrologic and erosion responses of the watershed. ACTMO has been tested using data from 2-year field experiments with the pesticide carbofarum with satisfactory results. This model does not need extensive calibration.

2.5.5 HSPF

The most extensively used general hydrologic and water quality model is HSPF, the Hydrologic Simulation Program Fortran (Barnwell and Johanson, 1981; Donigian et al., 1983). Its hydrology component is a modification of the Stanford Watershed Model (discussed in Section 2.2.3). HSPF is a general-purpose model that simulates the movement of dissolved oxygen, organic matter, pesticides, nutrients, salts, bacteria, sediment, and plankton from the land surface through channel and reservoir systems, including groundwater (DeCoursey, 1985). The biggest drawback to its use is that many of the hydrologic parameters are empirical, requiring several years of record to obtain parameter values.

HSPF is a comprehensive package for simulation of watershed hydrology and water quality developed by the EPA (Barnwell and Johanson, 1981; Donigian et al., 1983). It is a large model requiring large computer storage and considerable effort in preparation of data input. It contains three application moduli, PERLND, IMPLND, and RCHRES, and five utility moduli: COPY, PLTGEN, DISPLY, DURANL, and GENER. It basically simulates on a lumped parameter basis, and the magnitude of parameters must be determined by calibration.

The module PERLND simulates a pervious land segment with homogeneous climate and hydrologic characteristics. Overland flow and erosion are modeled by SWMM and NEGEV models, respectively (Negev, 1967). The module IMPLND is designed to simulate impervious land segments with no infiltration. The module RCHRES simulates the processes that occur in the single reach of an open channel or completely mixed lake. The HSPF system has been in public use since 1980 and it is available from the EPA. Donigian (1986) gives more details on HSPF.

2.5.6 CREAMS

The CREAMS model (Chemical, Runoff and Erosion from Agricultural Management Systems) has been developed by scientists in the Science and Education Administration-Agricultural Research (SEA-AR) of the USDA and consists of three major components of hydrology, erosion-sedimentation, and chemistry, as shown in Figure 2.17 (Knisel, 1980). The hydrology component estimates runoff either by the SCS Curve Number Model or by the Green and Ampt infiltration equation, depending upon the availability of data.

The CREAMS model is comprehensive, powerful, and well documented. It is essentially suitable for making relative comparisons of pollutant loads from alternative management practices. Because it computes transport of a range of particle sizes, it is capable of predicting selective transport and enrichment of fine particles in sediment, and therefore should provide a useful basis for predicting sorbed chemical transport. The erosion component considers the processes of detachment, transportation, and deposition. It assumes that sediment load is controlled either by the losses

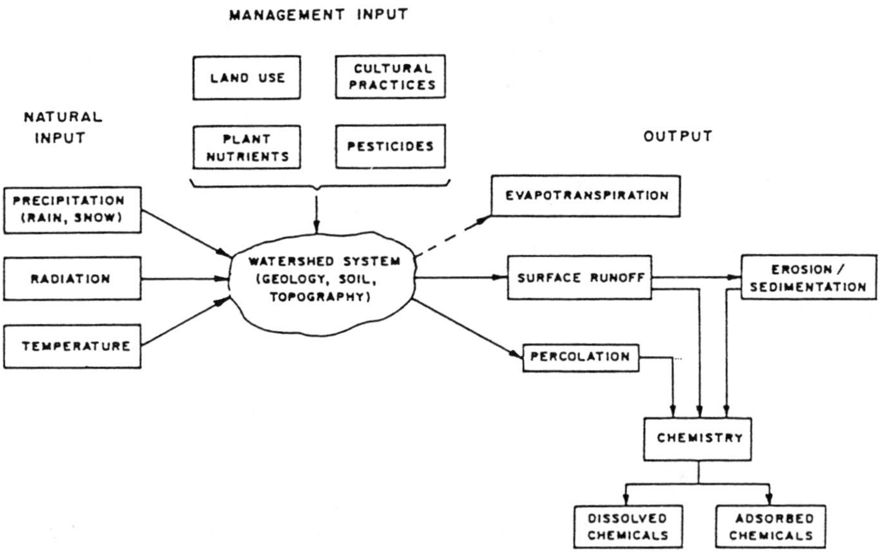

FIGURE 2.17. Flow chart of CREAMS model. (From Knisel, 1980. Report No. 26. U.S. Department of Agriculture, Washington, D.C. With permission.)

at transport capacity or by the amount of sediment available for transport. Detachment is described by a modification of the USLE for a single storm event. Transport capacity of overland and channel flow is derived from Yalin's sediment transport equation (Knisel, 1980).

The basic concepts of the nutrient component are as follow:

1. Nitrogen and phosphorus attached to soil particles are lost with the sediment.
2. Soluble nitrogen and phosphorus are lost with the surface runoff.
3. Soil nitrate is lost by leaching with percolating water, by denitrification, or by plant uptake.

The pesticide component estimates pesticide concentration in runoff, and the total mass of pesticide carried from the field for each storm during the period of interest. Pesticide in runoff is partitioned between the solution and the sediment phase using a linear isotherm model.

The major drawbacks of CREAMS are its complexity, intensive data requirements, and its reliance on modified USLE relationships and parameters. This degree of empiricism employed in the CREAMS model makes it useful for planning purposes and immediate application to field conditions, but limits its use for research in the physical processes causing erosion.

The CREAMS model has been used in many parts of the world with varying degrees of success, some of which are reviewed below.

The CREAMS model was successfully incorporated into an economic analysis designed to compare net return, soil loss, nitrogen, phosphorus, and pesticide losses in runoff resulting from agricultural cropping activities in a dairy farm in Pennsylvania (Crowder et al., 1985). CREAMS provided a useful method for estimating the effectiveness of agricultural management practices for controlling chemical losses and water quality.

Lorber and Mulkey (1982) used the CREAMS model to predict the average concentration of the herbicide hexazinone in stormflow from small forested watersheds in Georgia. CREAMS accurately predicted herbicide concentration in the initial stormflow events that contained the highest concentrations, but underestimated for subsequent stormflows. Subsurface movement of residues by interflow and deep leaching losses were not simulated in the version of CREAMS used. One of the limitations of CREAMS in simulating herbicide movement is the assumption that the infiltrated water and its herbicide content (dissolved herbicide) is unavailable for subsequent stormflow, while it is very likely that this infiltrated dissolved chemical will in part return to the surface flow further down the slope.

Sweeney et al. (1985) compared the CREAMS model prediction of nitrogen loss with the measured nitrogen loss by runoff and leaching from tomato and corn fields in Florida. Soil type appeared to influence the ability of CREAMS to predict seasonal trends and treatment influences. Model prediction of nitrogen loss from sandy soils disagreed with measured values, but more satisfactory prediction was obtained for higher clay content soils. The simulation ability of CREAMS for predicting nitrogen leaching loss needs improvement.

Although CREAMS is said to need no calibration, adjustment of input parameters led to a better agreement of predicted and observed values (Sweeney et al., 1985). The same conclusion with regard to model calibration was reached by Lorber and Mulkey (1982).

Knisel et al. (1985) found the comparison between observed runoff and that simulated by CREAMS without calibration generally good except in areas where snowmelt is a significant part of annual runoff. The hydrology component of the original CREAMS model contains a snowmelt element, but it does not include an ability to deal with seasonally frozen soils. The hydrology component of CREAMS was modified to represent frozen soil conditions during spring snowmelt periods without requiring additional input, and this modified CREAMS was tested at six locations where observed data were available (Knisel et al., 1985). Model simulations with and without the modification were compared with observed runoff. The modified model resulted in greater estimates of snowmelt runoff than the original model. Annual runoff simulated with the modified model compared very well with observed runoff. Simulated percolation through the root zone was less with the revised model on permeable soils than with the original model, in agreement with field data.

Nyhan and Lane (1982) and Hakonson et al. (1986) used CREAMS to predict water balance and soil loss and contaminant transport in the highly disturbed trench cap used for covering shallow land burial of radioactive wastes in the arid regions at Los Alamos, New Mexico under a variety of plant cover conditions. The use of CREAMS in arid/semiarid environments and waste management extends the model beyond its capabilities, primarily because data describing these conditions are not readily available. Animal interactions, which are not directly represented in the model, are important in altering water balance, erosion, and nutrient cycling. Until significant advances are made in our understanding and ability to mathematically describe processes in arid ecosystems, models such as CREAMS will be somewhat limited in their utility in such environments.

Rudra et al. (1985) applied the CREAMS model at the plot scale in arid/semiarid environments of southern Ontario under a variety of plant cover and soil conditions. Results showed wide discrepancies between the computed and observed runoff, soil loss, and sediment and nutrient load for individual events. Modifications were then made in the hydrology and erosion submodels of CREAMS but the overall performance of the modified submodel was not satisfactory.

The CREAMS erosion model has been evaluated for a range of soils in the semitropical region of Queensland, Australia (Loch et al., 1989; Silburn and Loch, 1990). Reasonable predictions of sediment yield were obtained for all the soils using values for flow hydraulic and soil detachment parameters derived from runoff and sediment data of rainulator plot experiments. Silburn et al. (1990) used the CREAMS model to investigate the effects of land management on sediment yield and export of fine sediment in the same region of Australia. The CREAMS model proved to be useful for this kind of study.

2.5.7 CREAMS-WT

CREAMS-WT (WT, water table) is a modification of the CREAMS model to improve the model conceptually and to enable it to represent the low nutrient sorption capacity of sandy soils of costal plains (Heatwole et al., 1987). The nutrient submodel of CREAMS has been modified to represent more realistically the nutrient dynamics of storage basins with flat topography, high water table, and sandy soil in which most of the nutrient loss is in soluble form. Some modification has also been made in the hydrologic submodel of the CREAMS model to be able to determine explicitly the number of days when saturated conditions exist in the upper profile based on the level of the water table (Heatwole et al., 1988). This modification is in the hydrology model but its effect is seen in the nutrient model when denitrification time is estimated. Also added to the model is the capacity to specify the concentration of nutrients in the rainfall. For validation two watersheds were simulated without calibration. Predicted annual flow-weighted concentrations of total nitrogen and total phosphorus compared closely with observed values for both sites, correctly reflecting the response of the two widely different cases of nutrient loading (Heatwole et al., 1988).

2.5.8 GLEAMS

GLEAMS (Groundwater Loading Effects of Agricultural Management Systems) is an extension of the CREAMS model to include a component for vertical flux of pesticides and ground-water loading effect (Leonard et al., 1987). Although its name indicates that it is primarily a groundwater loading model, GLEAMS is mainly concerned with surface runoff and erosion processes. GLEAMS uses a fixed time step of 1 day but it has an event-based erosion component. It allows modeling of crop rotation of a specific cycle length with different erosion and pesticide parameters

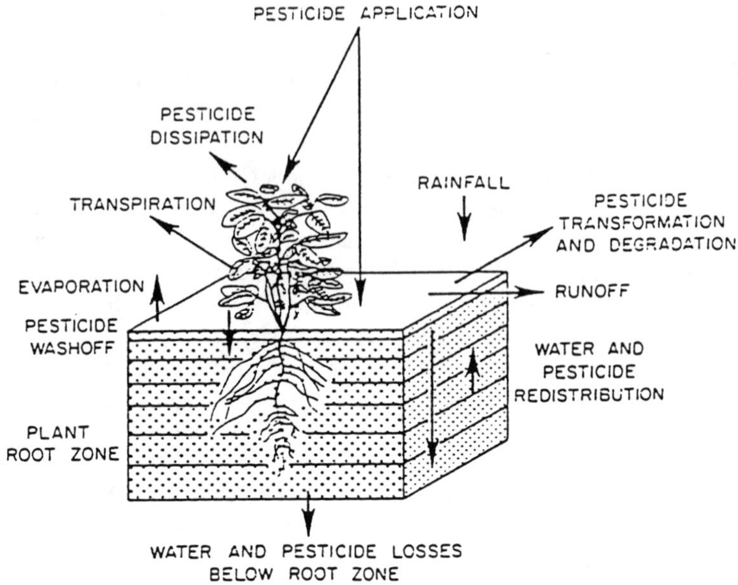

FIGURE 2.18. The physical system and processes represented in GLEAMS. (From Leonard et al., 1987. *Trans. ASAE* 30(5):1403–1418. With permission.)

used for each year. The physical system and processes represented in GLEAMS are shown in Figure 2.18.

GLEAMS consists of hydrology, erosion, and pesticide components. Because a layered soil system is needed for routing pesticides through the root zone, hydrology option 1 of CREAMS with seven computational layers (Williams and Nicks, 1982) has been used in GLEAMS. This hydrologic option uses a modified version of the SCS Curve Number technique for infiltration, which includes evapotranspiration and percolation components to maintain continuous water balance. Runoff volume is the difference between precipitation and infiltration. Water balance computation and storage routing technique are the same as those in CREAMS.

The erosion component of this model is essentially the same as that in the CREAMS model, which uses Foster's overland flow and channel flow models (Foster et al., 1980; Foster, 1982). The only significant change is the calculation of sediment particle characteristics.

The basic concepts of the pesticide component of GLEAMS are the same as those in CREAMS with some additions that enable the estimation of movement within and through the root zone. Because the principal objective of the GLEAMS model was to simulate the effect of management practices on pesticide movement within and through the root zone, it was found essential to maintain sensitivity in the surface soil layer as well. Therefore, the pesticide concepts from CREAMS have been retained in the GLEAMS surface component for the user to be able to examine both surface and subsurface responses for all pesticides without requiring a model for surface and another model for subsurface responses. Due to complexities such as differences in the application methods of pesticides and herbicides and the differences in the mixing efficiency of tillage equipments, a good method of calculating surface layer thickness has not been developed for the GLEAMS model. Therefore the surface layer in this model is assumed to have a constant 1-cm thickness. The pesticide degradation and the pesticide runoff components of GLEAMS are essentially the same as those in CREAMS (Leonard and Wauchope, 1980) with only slight modifications.

The list of input files for each of the components of this model is quite extensive. A user manual provides adequate description of these input files and their format, but does not give enough information on model theory or program details. Most information needed is similar to that for CREAMS, for which documentation is well developed and comprehensive.

Both GLEAMS and CREAMS were developed as management tools to assess the impacts of different agricultural management practices. They were not intended to be used for the prediction of the exact pesticide and sediment loading in a given situation. Thus, like CREAMS, GLEAMS requires on-site calibration.

Some improvement in various components of the existing version of GLEAMS model is under way (Leonard et al., 1987). These improvements include addition of a new plant nutrient component different from that in CREAMS and a new climate generator, modification of pesticide component to consider degradation products, and the revision of the erosion component.

2.5.9 GAMESP

GAMESP is a model for evaluating the effects of agricultural management systems on erosion, sedimentation, and phosphorus yield (Rousseau et al., 1985). GAMESP consists of two major components of potential soil loss/sedimentation and phosphorus loss. It is the result of adding a phosphorus component to the GAMES model (described in Section 2.3.9). The conceptual approach of GAMESP involves discretization of agricultural watershed into land and stream cells of various size and shapes. This conceptual approach allows the GAMESP to view the watershed as a system of interconnected land and stream cells and model sediment and phosphorus transport by a series of interconnected land cells delivering sediment and phosphorus loading into a series of stream cells leading to the watershed outlet.

The potential soil loss/sedimentation component of GAMESP estimates the seasonal potential soil loss from each land cell, using a seasonal application of USLE, and the subsequent seasonal amount of sediment delivered to the adjacent downstream land cell and stream cell using a seasonal sediment delivery ratio. The seasonal sediment load at the outlet of the watershed is then the summation of all predicted cell-to-stream sediment loads. The phosphorus component of GAMESP estimates for each land cell the seasonal total phosphorus concentration of surface soil using a regression equation (if average total phosphorus concentration of surface soil is not known), the seasonal cell-to-cell phosphorus enrichment ratio and load, and the seasonal cell-to-stream phosphorus enrichment ratio and load. The resulting seasonal phosphorus load at the outlet of the watershed is calculated by summing up all predicted cell-to-stream phosphorus loads.

GAMESP needs calibration, and only after extensive field validation can it become a working tool useful as a guide in practical management of field situations.

2.5.10 WASCH

This research-oriented model of WASCH (Water, Sediment, and Chemical Transport) is developed for calculating water, sediment, and pesticide transport from a watershed for a storm event (Bruce et al., 1973). The water component describes the processes of rain falling on the watershed, its partition into effective and noneffective amounts in producing runoff, and the distribution of the effective portion to form streamflow. The sediment component is built on the rill and interrill erosion concepts of Foster and Meyer (1975). Rill and interrill erosion are conceptually distinguished, which allows similar partitioning of associated pesticides. The interrill erosion is primarily due to rainfall impact. Dislodged material is transported into rills by overland flow or direct splash. Sediment transport capacity is the sum of rill and interrill concentration functions. The product of sediment concentration and water discharge is the mass movement of sediment for the time interval. Summation of the amounts across the total storm duration gives the load of sediment for the storm. The pesticide component of this model makes no attempt to compute water- and sediment-transported pesticides separately. Pesticide concentration is computed for the total runoff of water plus sediment. The model was calibrated and applied to a 2.7-ha watershed with rolling topography for four short-duration high-intensity storms. Good agreement between predicted and measured results was obtained for catchments similar to those used in calibration.

2.5.11 AGNPS

The Agricultural Nonpoint Source (AGNPS) model is a computer simulation model developed by Young et al. (1987) to analyze the water quality of runoff from Minnesota watersheds. The model predicts runoff volume and peak rate, eroded and delivered sediment, and nitrogen, phosphorus, and chemical concentrations in the runoff and in the sediment for single storm events for all points in the watershed. Watersheds examined by the model must be segmented into square working areas called cells. This allows the model to provide water-quality information for specific locations within the analyzed watersheds. Groupings of the cells make up numerous subwatersheds that then could be separately analyzed to provide further information on individual watershed sections. This model is capable of analyzing pollutant loads from feedlots and of investigating the effects of implementing various conservation practices, including terracing. It is also capable of providing water-quality characteristics output at intermediate points throughout the watershed network.

The basic components of the model are hydrology, erosion sediment transport, and transport of nitrogen, phosphorus, and oxygen-demanding chemicals. In the hydrology portion of the model, calculations are made for runoff volume and peak concentrated flow. For estimating direct runoff from storm water, the model uses the SCS Curve Number Equation. The erosion portion calculates total upland erosion, total channel erosion, and the breakdown of these two into five particle size classes (clay, silt, small aggregates, large aggregates, and sand) for each of the cells. Sediment transport is also calculated for each of the cells in the five particle classes as well as the total. The pollutant transport portion is subdivided into one part handling soluble pollutants and another part handling sediment-attached pollutants. AGNPS has two versions of AGNPS-I suitable for large watersheds (500 to 23,000 acres) and AGNPS-PC or AGNPS-II for small watersheds of 2.5 to 8000 acres (Young et al., 1985; USDA-ARS, 1986).

2.5.12 SPNM

SPNM (Sediment, Phosphorus, Nitrogen Model) is a total sediment yield model developed by Williams (1980). An early version of the sediment routing model (Williams, 1975a) routed the sediment yield for an individual storm for each subwatershed to the watershed outlet considering the travel time and the median particle size. It included deposition and degradation components, but the deposition process was emphasized. This model was later refined (Williams and Hann, 1978) by replacing the median particle size with the entire particle size distribution, and a technique was developed to determine the deposition parameter for each routing reach instead of using one value for the entire watershed. Further refinement (Williams, 1978a) included the development of a new degradation component based on stream power. This latest version of the sediment routing model was combined with the modified USLE (MUSLE) (Williams, 1975b) and nutrient loading function to form a water-quality model called SPNM. SPNM is written in the form of a problem-oriented computer language for solving water-quality problems on large watersheds (larger than 2500 km^2).

2.5.13 EPARRB

EPARRB (Environmental Pollution Assessment-Erosion Sediment and Rural Runoff) is a model suitable for large, single-catchment, nonurban watersheds. It is a stochastic model using a random number generator to introduce probabilistic highly variable conditions (True, 1972, 1976). A deterministic system can be obtained by using mean values. It can be used as a single storm event model for erosion and sedimentation with the option of monthly input and output. No particular hydrologic model is represented and the USLE is used for erosion and sediment loss. It was intended for concurrent use with the EPA Urban, Commercial, and Industrial Runoff model (EPAURA). Loadings from both models are summed along with point source loadings to determine total core city, urban, and rural loadings to water bodies (EPATLC). EPARRB does not consider rainfall and runoff hydrology; only empirical relationships of erosion resulting from rainfall and runoff are considered. Sediment transport is also derived from observational data rather than simulation of individual processes. Chemical constituents represented are nitrogen, phosphorus, and animal waste. EPARRB is useful for "first cut" estimations of nonpoint source contributions to receiving water bodies. It has poor temporal and spatial resolution as well as a lack of process detail (True, 1976).

2.5.14 SWMM and STORM

The EPA Storm-Water Management Model (SWMM) (Metcalf and Eddy, Inc., 1971; Huber et al., 1981) and the U.S. Army Corps of Engineers Storage Treatment, Overflow Runoff Model (STORM) (U.S. Army Corps of Engineers, 1977) are the two most commonly used deterministic urban models. SWMM continuously simulates runoff quality and quantity from pervious and impervious areas, sedimentation and scour, pollutant routing in sewers, and receiving water quality, storage, and treatment. STORM, while similar to SWMM, is capable of continuous simulation with hourly time steps whereas the original version of SWMM is an event model. SWMM is an empirical model containing a number of parameters that have no physical basis (Osman, 1987). It considers suspended solids to be the primary indicator of pollutant wash-off with an exponentially decaying form. It relies on accurate estimation of a large number of empirical coefficients and needs extensive calibration. STORM uses the SCS Curve Number Equation as its hydrologic component, while SWMM uses the Green and Ampt infiltration equation and Manning's overland flow equation and continuity. A simplified version of SWMM has also been developed by Metcalf and Eddy, Inc. (1971) for medium to large urban watersheds.

2.5.15 PLIERS

The Pesticide Losses In Erosion and Runoff Simulator (PLIERS; Kenimer et al., 1989) is a field-scale model for predicting the loss of pesticides in surface runoff, in sediment, and by degradation and volatilization. It also accounts for the wash-off of pesticides from surface cover and the movement of pesticides between the adsorbed and dissolved phases. Hydrologic data required by PLIERS are provided by a watershed model called FESHM, which contains an enhanced sediment algorithm that predicts the detachment and transport of individual sediment size classes (Ross et al., 1978). PLIERS uses first-order rate equations to describe pesticide degradation, volatilization, and wash-off. The adsorption of pesticides to individual particle size classes is estimated using the Freundlich Equation (described in Section 2.5.13). A regression equation has been developed to determine the Freundlich K_D value as a function of the mean diameter of a soil sediment size class (Kenimer et al., 1989). PLIERS estimates degradation and volatilization losses of pesticides on a daily basis and assumes that these losses are independent of soil pH, soil moisture content, and ambient conditions. Such assumptions impose limitations on the applicability of the model and could become a major source of error.

PLIERS was validated using atrazine and 2,4-D losses from rainfall simulator studies of field plots with different tillage types and residue levels. Agreement between predicted and observed data was satisfactory for both pesticides. PLIERS shows some potential as a flexible planning tool and it is capable of directly reading in hydrologic data or using any comprehensive watershed model as its submodel.

2.5.16 PRM

The Pesticide Runoff Model (PRM) is a simple screening model that can be used to evaluate pesticide losses for a wide range of field conditions and pesticide characteristics (Haith, 1980). The model is not necessarily better or more accurate than previous models, but it is designed for use in the field situations where pesticide runoff data are not available. Because the data requirement is modest and the model consists of several straightforward mathematical relationships it should be comprehensible to most potential users.

This model is based on a mass balance of the surface 1 centimeter of soil. It is assumed that pesticides below this depth are not available for runoff loss. When the pesticide is surface applied, all of the chemical is initially available for runoff. With a soil-incorporated pesticide, only the portion of the chemical that remains in the 1-cm surface layer is available. As with previous models, pesticide levels in the soil are considered to decay exponentially with time. If the first rainstorm occurs t days after pesticide application, the level of pesticide in the surface 1 cm (in g ha^{-1}) is given by

$$P_t = P_0 \exp\left(-\alpha t\right) \tag{46}$$

where P_0 is the initial pesticide content (g ha^{-1}) of the surface soil immediately following application and α is the pesticide decay rate (day^{-1}).

The total pesticide P_t can be divided into adsorbed (A_t) and dissolved (D_t) based on a single-valued linear equilibrium adsorption isotherm:

$$P_t = A_t + D_t \tag{47}$$

The solid-phase loss (PX_t) and dissolved loss (PQ_a) of pesticide (in g ha^{-1}) is calculated using the following equation:

$$PX_t = \left[X_t / 100\delta\right] A_t \tag{48}$$

$$PQ_a = \left[Q_a / R_t\right] D_t \tag{49}$$

where X_t and Q_a are soil loss (in tons ha^{-1}) and runoff (in cm) respectively, R_t is the rainfall (in cm), and δ is soil bulk density (in g cm^{-3}).

Although Haith's pesticide runoff model is not restricted to use of any specific hydrologic or soil loss model, the CNE (SCS Curve Number Equation; see Section 2.2.2) and William's MUSLE (described in Section 2.3.2) have been used for the estimation of runoff and soil loss in the model. Both curve number and the USLE are applicable only to rainfall, so the model cannot be used to estimate pesticide losses in runoff from snowmelt. The predicted dissolved and solid-phase losses are "edge-of-field" values and provide only upper limits of pesticide input to surface waters. This model is not suitable for isolated rainstorms, but its performance for averaged long period predictions of 3 years or more has been reasonably good (Haith, 1980).

Model validation and testing show that the pesticide predictions varies considerably among events. The model is not suitable for predicting pesticide losses from a single isolated rainstorm. The model performs better for the estimation of average or seasonal losses.

This model was later combined with a weather generator to permit long-term simulations, as shown in Figure 2.19, and used for a 100-year simulation of runoff losses of atrazine and carbofuran from different soil types and under various climatic conditions (Haith, 1986).

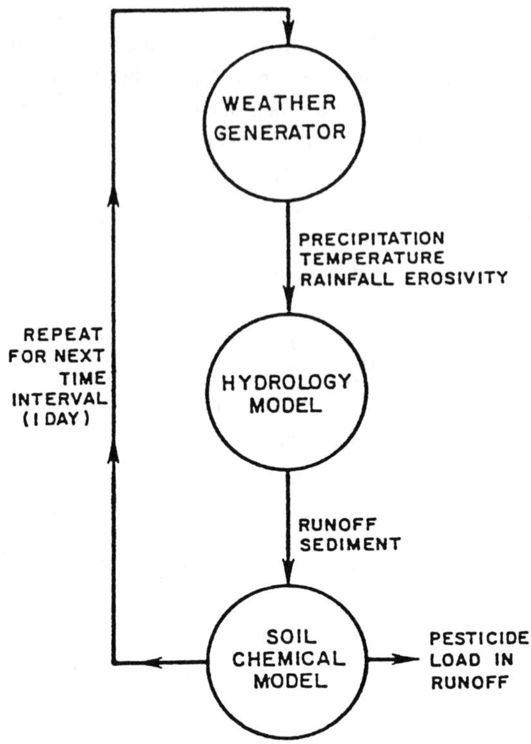

FIGURE 2.19. PTR model structure. (From Haith, 1986. *J. Environ. Qual.* 15(1):5–8. With permission.)

2.5.17 CNS

The Cornell Nutrient Simulation (CNS) model consists of three basic components of daily soil moisture, daily soil loss, and monthly nitrogen and phosphorus losses (Haith and Loehr, 1979). For its hydrologic component CNS relies on the SCS Curve Number Equation. A daily soil moisture inventory is maintained for the top 30 cm of soil, assumed to be homogeneous. Soil loss is estimated by Onstad and Foster's modification of the USLE (Onstad and Foster, 1975). Daily runoff, percolation, and soil loss are summed for 1 month and used as input to the nutrient and pesticide model (CNS and CPM). Monthly inventories of mass balances are computed for adsorbed and dissolved nutrients in the top 30 cm of soil. Denitrification, ammonia volatilization, and ammonium fixation are not modeled explicitly. Instead a flat 25% of fertilizer nitrogen applications is subtracted to account for these losses. Soil phosphorus is divided into available and fixed inorganic forms, and available phosphorus subdivided to dissolved and adsorbed components based on a linear equilibrium isotherm. When tested in the field, simulated and measured data did not match very well.

2.5.18 CPM

The Cornell Pesticide Model (CPM) consists of five components concerned with soil temperature and snowmelt, hydrology, soil loss, pesticide movement, and statistics. Runoff volume is worked out in the same way as by the CNS, and the sediment submodel is similar to that of the CNS (Haith and Loehr, 1979). The pesticide component consists of three parts: degradation and volatilization, downward displacement of pesticide, and pesticide in overland flow. The model uses the SCS Curve Number Equation or the Green and Ampt (1911) infiltration equation to determine the volume of surface runoff (Novotny, 1986).

2.5.19 NONPT

This is a version of the USDAHL-70 model prepared by the Agricultural Experimental Station of the University of Maryland (Holtan and Lopez, 1971; Holtan, 1979) that is capable of modeling nitrogen and phosphorus transport by runoff. Phosphorus is transported to sediment load and the amount of adsorbed or desorbed phosphorus is computed using an equation given by Novotny et al. (1978). Like ACTMO, this model does not require extensive calibration.

2.5.20 SWAM

SWAM (Small Watershed Model) is a comprehensive model for simulating the response of watersheds less than 25 km^2 in size by routing water, sediment, and chemicals from all sources through channel reservoir and ground waters (DeCoursey, 1982). It uses CREAMS to simulate the response from all land use areas. SWAM has been developed to study off-site benefits of nonpoint routing water, sediment, nutrients, and pesticides from each of the CREAMS elements. Biological, physical, and chemical features of most processes are simulated.

2.5.21 MRI-NSM

The Midwest Research Institute Nonpoint Source Model (MRI-NSM) is a deterministic model applicable to irrigated and nonirrigated small agricultural watersheds. Inputs are usually annual averages and it uses the USLE for sediment erosion prediction. Hydrology is not considered directly. The constituents of the model are sediment-attached nitrogen, phosphorus, organic matter, pesticides, salinity, and heavy metals. It is easy to use with loading function documentation reports available through the EPA. Output is an average 30-day maximum daily loading of pollutant from each land use category. The technique permits little temporal or spatial variation or process detail (McElroy et al., 1976).

2.5.22 UTM

The Unified Transport Model (UTM) has been developed by Oak Ridge National Laboratory. The UTM primarily describes the movement of heavy metals, but it can simulate the atmosphere-soil-plant-water system, plant growth, soil chemical exchange of heavy metals, and soluble uptake by vegetation (Huff et al., 1977). The

UTM transport mechanism is based primarily on the Stanford Watershed Model (Crawford and Linsley, 1966).

2.6 COMPARING SOME OF THE MORE WIDELY USED MODELS

2.6.1 Introduction

Not much work has been done on comparing the performance of various nonpoint source pollution models under different conditions. Although a large number of such models have been developed during the last two decades, literature on their comparative studies is scarce. However, because such comparative studies are capable of playing an important role for the potential users of such models, a review of the limited number of papers available on this subject is given below.

2.6.2 Comparative Evaluation of ARM, CPS, and CREAMS Models

Lorber and Mulkey (1982) evaluated the performance of ARM, CPS, and CREAMS by applying them to estimate field-scale runoff losses of two pesticides applied to a cotton field and a watershed. All three models were used for a 5-year simulation under identical conditions. The main differences among these models are in hydrologic and erosion components. The hydrologic components of these models vary from the classical water balance of the Stanford Watershed Model to the more theoretically based Green and Ampt infiltration equation. The descriptions of erosion mechanics vary from modification of the USLE for total erosion to the bed-load transport equation for rill erosion (Negev, 1967; Yalin, 1963).

In all the models pesticide degradation processes are assumed to obey first-order kinetics. Time- and space-variable first-order processes are permissible only in the ARM model.

The major difference between ARM and the others is the detail with which storms are represented: storm total only by CPS and CREAMS, time steps within storm by ARM. Another difference is that the ARM model explicitly simulates interflow and associated pesticide transport while other models do not.

All models simulated monthly total runoff and peak runoff quite well. The CPS and ARM models simulated sediment loss equally well but CREAMS underpredicted loss, especially in peak months.

CREAMS predicted 25% less erosion loss of adsorbed pesticides than the other two, perhaps because CREAMS relies entirely on an enrichment ratio, while the other two models partition chemicals between runoff sediment and water. For all three models the correlation coefficients were higher for runoff and sediment predictions than for pesticide loss.

It is claimed that the CPS and CREAMS models do not need calibration, but the results were improved by some calibration and reassignment of initial parameters estimates.

2.6.3 Comparative Study of Six Hydrologic Models

Six hydrologic models for estimating runoff peak rates were applied to Florida's Flatwoods watersheds under small to moderate rainfall events (Capece et al., 1988).

Watersheds were of gentle slope, highly permeable sandy soils, with high water tables and wetlands. These characteristics are unlike those of watersheds that served as models for the development of most peak runoff rate prediction methods. Five watersheds were used in this study, ranging in size from 8 to 1450 ha of improved and unimproved pastures. The following six runoff peak rate estimating techniques were compared in this study, beginning with the most empirical, and progressing toward more theoretically based methods:

1. Cypress Creek formula (Stephens and Mills, 1965)
2. CREAMS hydrologic equation (Knisel, 1980)
3. SCS graphical method (USDA-SCS, 1980)
4. SCS chart method (USDA-SCS, 1980)
5. SCS triangular hydrograph or unit hydrograph method (USDA-SCS, 1980)
6. South Florida water management district model (SFWMD, 1983; Higgins, 1976)

Among the methods examined the SFWMD performed best, while the CREAMS hydrologic equation yielded the least accurate estimates. The ranking of all six methods, based on overall mean percentage error of estimate, was as follows: CREAMS, 2200%; SCS chart, 1120%; Cypress Creek formula, 490%; SCS triangular hydrograph, 240%; and SFWMD, 36%. The CREAMS, SCS graphical, and SCS triangular hydrograph methods were then modified to reflect observed data better (Capece et al., 1988). However, the results obtained from these modified methods no longer represented independent evaluation of the methods but simply reflected the regression fit to the data, and not surprisingly produced much more accurate prediction than original equations or graphs.

Results of this study suggest that more accurate estimates of runoff peak rate can be expected as models progress from the empirical to the more physically based. However, when empirical models are tailored to specific watershed conditions, results may be similar to or even better than those from more physically based methods. As watershed conditions change, or changes are anticipated, physically based models again become more reliable than empirical techniques, because there is some basis for considering the consequences of change.

2.6.4 Comparing Runoff-Simulating Performance of the USDAHL and CREAMS Models

Pathak et al. (1984) compared the performance of the USDAHL-74 and CREAMS models in simulating monthly and annual runoff from four grassland watersheds ranging in area from 6 to 80 ha. USDAHL simulated monthly runoff more accurately than CREAMS on all watersheds. Both models were more accurate for small watersheds with relatively homogeneous soils than for large ones with heterogeneous soils. Compared with the observed runoff, the CREAMS model underestimated monthly runoff during cool seasons and overestimated during warm seasons. This simulation error was especially serious on the large watersheds.

2.6.5 Comparing the HILLS Model with the Hydrologic Component of the ANSWERS Model

The performances of the two physically based rainfall-runoff models (HILLS and ANSWERS) were compared using data obtained from two small but fully instrumented catchments in the wheatbelt region of western Australia, where contour banks had been in place for over 25 years (Townley and Bligh, 1990). The two models tested offered different advantages but shared common problems. ANSWERS proved to be relatively easy to set up for individual contour bays and capable of simulating a two-dimensional network that covered the plan areas of small catchments. Its algorithms were found to be simple and it was capable of simulating complex events with good accuracy. HILLS, on the other hand, offered the advantage of being able to predict shallow groundwater levels as well as surface flows. However, the model was much harder to calibrate and appeared to be less robust, in the sense of its ability to converge over a wide range of parameter values. Both models in the forms implemented in this study proved to be time consuming to tailor for a new situation. Both models require user-friendly graphic interfaces if they are going to be used by field officers for design or evaluation of erosion control measures. Both models also showed sensitivity to antecedent moisture conditions to the extent that their use in practice would need to be coupled to some other long-term soil moisture accounting model.

2.6.6 Comparing Chemical Process Components of Major Models

A summary of the pesticide process components of the major models is given in Table 2.2 (from Mulkey et al., 1986). The importance of the differences outlined here are difficult to assess. In most cases the properties of the specific chemical, its application procedure, and its formulation determines the relative importance of each process.

Table 2.2 Pesticide Process Representation in Current Pesticide Runoff Models

Model	Transport Mechanisms	Degradation	Soil-Water Partitioning	Plant Processes	Other Options
PRS	Dissolved in surface runoff, sorbed to eroded soil	First order, lumped decay rate	Reversible, instantaneous, equilibrium, single-value linear isotherm	Foliar washoff and decay	Enrichment ratios
CREAMS	Dissolved in surface runoff, sorbed to eroded soil	First order, lumped decay rate	Reversible, instantaneous, equilibrium, single-value linear isotherm	Foliar washoff and decay	Enrichment ratios
CPS	Dissolved in surface runoff, sorbed to eroded soil	First order, lumped decay rate	Reversible, instantaneous, equilibrium, single-value linear isotherm	None	None
ARM II	Dissolved in surface runoff Dissolved in interflow Sorbed to eroded soil	Time-phased series of first order, lumped decay rates	Reversible and irreversible, single and nonsingle valued, Freundich isotherm	None	None
HSPF	Dissolved in surface runoff Dissolved in interflow Sorbed to eroded soil	Time-phased series of first order, lumped decay rates	Same as ARM II with added option for sorption/desorption kinetics	None	None

After Mulkey et al. (1986).

2.7 LIST OF SYMBOLS

a	Gross detachability of the soil to rainfall, $M\ L^{-3}$
a_d	Gross redetachability of deposited sediment to rainfall, $M\ L^{-3}$
A	Average annual soil loss, $M\ L^{-2}$
A_e	Event soil loss, $M\ L^{-2}$
A_i	Area increment, L^2
A_s	Seasonal sediment yield delivered from one point to the other or to a stream, $M\ L^{-2}$
A_t	Adsorbed pesticide content, $M\ L^{-2}$
A_P	Labile phosphorus content of sediment, $M\ M^{-1}$
A_u	Labile phosphorus pool prior to runoff, $M\ M^{-1}$
c	Sediment concentration, $M\ L^{-3}$
c_i	Sediment concentration in size class i, $M\ L^{-3}$
c_t	Sediment concentration at the transport limit, $M\ L^{-3}$
C	Cropping management factor
C_a	Average concentration of pollutant n in the surface layer of the soil, $M\ M^{-1}$
C_c	Crop cover factor (a combination of C and P factors in the USLE)
C_e	Interrill canopy cover coefficient
C_f	Water surface profile shape factor
C_p	Total pesticide concentration in runoff, $M\ L^{-3}$
C_s	Solution-phase pesticide concentration, $M\ L^{-3}$
C_{sp}	Pesticide concentration in 0 to 1 cm of top soil, $M\ M^{-1}$
d_i	Rate of sediment deposition of sediment in size class i, $M\ L^{-2}T^{-1}$
D	Depth of water in overland flow, L
D_{cf}	Net detachment or deposition in the channel, $M\ L^{-2}\ T^{-1}$
D_i	Interrill erosion rate, $M\ L^{-2}\ T^{-1}$
D_f	Flow detachment rate, $M\ L^{-2}\ T^{-1}$
D_r	Rill erosion rate, $M\ L^{-2}\ T^{-1}$
D_t	Dissolved pesticide content, $M\ L^{-2}$
DR	Seasonal delivery ratio between two selected points
D_s	Soil detachability index defined as the mass of soil detached per unit of rainfall energy, $L^{-2}\ T^2$
e	Rate of rainfall detachment per unit area of soil, $M\ L^{-2}\ T^{-1}$
e_i	Rate of rainfall detachment of soil in size class i (per unit area), $M\ L^{-2}\ T^{-1}$
e_{di}	Rate of rainfall redetachment of deposited sediment in size class i (per unit area), $M\ L^{-2}\ T^{-1}$
E	Kinetic energy of rainfall per unit area, $M\ T^{-2}$
E_a	Daily actual evapotranspiration, L
E_p	Daily potential evapotranspiration, L
ER_n	Enrichment ratio of pollutant n
E_s	Soil loss from a field (the product of soil loss per unit area and the field area), M

f	Ratio of mean particle density of soil to that of eroded sediment
F	Fraction of excess stream power utilized in entrainment and reentrainment
F_s	Fraction of pesticide lost in sediment phase
F_w	Fraction of pesticide lost in water phase
g	Acceleration due to gravity, $L\ T^{-2}$
G	Transport capacity of overland flow, $M\ L^{-2}$
G_e	Factor for the effect of ground cover on interrill erosion
G_f	Rate of sediment movement in flow, $M\ T^{-1}$
G_s	Ground cover effect coefficient
H	Fractional coverage of the original soil matrix by a layer of deposited sediment
H_c	Hydraulic coefficient
i	Subscript: refers to a particular range of sediment settling velocity (or size)
I	Infiltration rate, $L\ T^{-1}$
I_a	Initial abstraction, L
I_r	Typical value for intensity of erosive rain, $L\ T^{-1}$
I_t	Total number of settling velocity classes of equal mass interval
J	Specific energy of entrainment, $L^2\ T^{-2}$
k	Decay rate of labile phosphorus
k_1	Constant involving length, slope, and roughness of the plane
K	Soil erodibility factor in the USLE
K_1	Kinematic rate coefficient for pesticide adsorption, T^{-1}
K_2	Kinematic rate coefficient for pesticide desorption, T^{-1}
K_c	Coefficient for slope and roughness of the plane
K_D	Adsorption partition coefficient for pesticides, $M^{-1}\ L^3$
$\overline{K}_D$	Average adsorption partition coefficient for pesticides, $M^{-1}\ L^3$
K_i	Interrill erodibility coefficient
K_L	Langmuir coefficient, $M^{-1}\ L^3$
K_r	Rill erodibility coefficient
L	Length of slope, L
L_d	Length of surface drainage flow path between two locations, L
L_e	Effective rainfall rate, $M\ T^{-1}$
L_n	Load of pollutant-delivered to the receiving water body, $M\ L^{-1}$
M	Rate of sediment transport, $M\ L^{-1}\ T^{-1}$
M^*	Sediment load normalized to transport capacity
M_i/M_t	Ratio of mass of soil in size class i of the original soil matrix to M_t, where $M_t = \Sigma\ M_i$
M_{di}/M_{dt}	Ratio of mass of soil in size class i in the deposited layer to M_{dt}, where $M_{dt} = \Sigma\ M_{di}$
n	Manning's surface roughness coefficient
p_i	The probability of finding a grain of particular size in the stream bed
P	Erosion-control practice factor in the USLE
P_a	Available phosphorus content of soil, $M\ M^{-1}$
P_d	Cumulative phosphorus desorbed, $M\ T^{-1}$

P_i	Phosphorus content of particle of size class i, $M\ M^{-1}$
P_n	Potency factor of pollutant n
P_0	Initial pesticide content of soil, $M\ L^{-2}$
P_P	Total amount of sediment-bound phosphorus transported, $M\ L^{-2}$
PQ_a	Dissolved pesticide loss, $M\ L^{-2}$
P_r	Percentage rainfall contributing to permanent interception and stemflow
P_{sb}	Sediment-bound phosphorus, $M\ M^{-1}$
P_t	Final pesticide content of soil, $M\ L^{-2}$
PX_t	Solid-phase pesticide loss, $M\ L^{-2}$
q	Volumetric water flux per unit width of plane, $L^2\ T^{-1}$
q_l	Lateral flow sediment load to channel, $M\ L^{-2}\ T^{-1}$
Q	Runoff rate per unit area of plane, $L\ T^{-1}$
Q_a	Actual runoff, L
Q_e	Event runoff volume, L^3
Q_m	Potential maximum depth of runoff, L
Q_p	Peak runoff rate, $L^3\ T^{-1}$
Q_u	Volume of overland flow, L^3
r	Factor of the lower zone water storage
r_i	Rate of entrainment of size class i sediment from the soil matrix, $M\ L^{-2}\ T^{-1}$
r_{ri}	Rate of reentrainment of deposited sediment of size class i, $M\ L^{-2}\ T^{-1}$
R	Rainfall erosivity, $M\ L^{-1}$
R_a	Annual rainfall, L
R_c	Critical depth of daily rainfall to cause runoff, L
R_d	Rainfall detachment rate, $M\ L^{-2}\ T^{-1}$
R_e	Excess rainfall rate, $L\ T^{-1}$
R_0	R_a/R_n, where R_n is the number of rain days in the year
R_r	Rainfall rate, $L\ T^{-1}$
R_u	Rainfall volume, L^3
R_s	Rill spacing (distance between rills), L
R_t	Total precipitation, L
S	Slope (sine of angle of inclination)
S_a	Amount of pesticides adsorbed to soil, $M\ M^{-1}$
S_d	Splash detachment, $M\ L^{-2}$
S_l	Load of sediment delivered to water body, $M\ T^{-1}$
S_m	Potential maximum retention
S_t	Storage per unit area of plane
T_c	Sediment transport capacity in the rill, $M\ L^{-1}T^{-1}$
T_{ce}	Sediment transport capacity at the end of the uniform slope, $M\ L^{-1}\ T^{-1}$
v_i	Settling velocity of sedimentary particles of size range i, $L\ T^{-1}$
V	Velocity of overland flow, $L\ T^{-1}$
V_a	Average velocity of overland flow between two locations, $L\ T^{-1}$
w	Computed rill width, L
WS	Water-to-soil ratio, $L^3\ M^{-1}$
x	Distance downslope, L

X	Distance along flow surface, L
X_t	Soil loss per unit area, $M\ L^{-2}$
ξ	The depositability of soil, $L\ T^{-1}$
ρ	The density of water, $M\ L^{-3}$
σ	Sediment density, $M\ L^{-3}$
τ_f	Flow shear stress, $M\ L^{-1}T^{-2}$
τ_c	Critical shear stress or the rill detachment threshold parameter, $M\ L^{-1}\ T^{-2}$
Ω	Stream power of overland flow, $M\ T^{-3}$
α	Pesticide decay rate, T^{-1}
δ	Soil bulk density, $M\ L^{-3}$
θ	Soil water content

2.8 REFERENCES

Akan, O. 1987. Pollutant washoff by overland flow. *J. Environ. Eng.* 113(4):811–823.

Anderson, E. A. and N. H. Crawford. 1964. The synthesis of continuous snowmelt runoff hydrographs on a digital computer. Technical Report No. 36. Department of Civil Engineering, Stanford University, Palo Alto, CA.

Bailey, G. W., R. R. Swank, Jr., and H. P. Nicholson. 1974. Predicting pesticide runoff from agricultural land: a conceptual model. *J. Environ. Qual.* 3:95–102.

Baker, J. L. 1980. Agricultural areas as nonpoint sources of pollution. In: Environmental Impact of Nonpoint Source Pollution. M. R. Overcash and J. M. Davidson (Eds.). Ann Arbor Science, Ann Arbor, MI.

Baker, J. L. and H. P. Johnson. 1978. Movement of pesticides and nutrients with water and sediment as affected by tillage practices: a field study. Off. Water Resour. Technol., Washington, D.C. NTIS No. PB 293145.

Baker, D. N., J. M. McKinion, and J. R. Lambert. 1983. GOSSYM: a simulator for cotton growth and yield. *Clemson Agric. Exp. Stn. Bull.* 1089.

Barnwell, T. O. and R. Johanson. 1981. HSPF: a comprehensive package for simulation of watershed hydrology and water quality. Proc. of the Seminar on Nonpoint Pollution Control Tools and Techniques for the Future. Interstate Commission on the Potomac River Basin, Rockville, MD.

Barnwell, T. O. and P. A. Krenkel. 1982. The use of water quality models in management decision making. *Water Sci. Technol.* 14:1095–1107.

Beasley, D. B. 1986. Distributed parameter hydrologic and water-quality modelling. In: *Agricultural Nonpoint Source Pollution: Model Selection and Application.* A. Giorgini and F. Zingales (Eds.). Elsevier, New York, pp. 345–362.

Beasley, D. B. and L. F. Huggins. 1980. ANSWERS (Areal Nonpoint Source Watershed Environment Simulation) — User's Manual. Department of Agricultural Engineering, Purdue University, West Lafayette, IN.

Beasley, D. B., L. F. Huggins, and E. J. Monke. 1980. ANSWERS: a model for watershed planning. *Trans. ASAE* 23:938–944.

Beasley, D. B., L. F. Huggins, and E. J. Monke. 1982. Modelling sediment yields from agricultural watersheds. *J. Soil Water Conserv.* 37:113–117.

Bogardi, I. and L. Duckstein. 1978. Input for stochastic control model of P loading. *Ecol. Model.* 4:173–195.

Boughton, W. C. 1987. Evaluating partial areas of watershed runoff. *J. Irrig. Drain. Eng.* 113:356–366.

Boughton, W. C. 1989. A review of the USDA-SCS Curve Number Method. *Aust. J. Soil Res.* 27:511-523.

Box, G. E. P. and G. M. Genkins. 1970. Time series analysis. Holden-Day, San Francisco, CA.

Brakensiek, D. L. and W. J. Rawls. 1983. Agricultural management effect on soil water processes. II. Green and Ampt parameters for crusting soils. *Trans. ASAE* 28:1753–1757.

Brakensiek, D. L., A. L. Heath, and G. H. Comer. 1966. Numerical techniques for small watershed flood routing. ARS41-113. Agricultural Research Service, U.S. Department of Agriculture, Washington, D.C.

Brenneman, L. G. and J. M. Laflen. 1982. Modelling sediment deposition behind corn residue. *Trans. ASAE* 25(4):1245–1250.

Bruce, R. R. et al. 1973. Water-Sediment-Chemical Effluent Prediction (WASCH Model). Southern Piedmont Conservation Research Center, Agricultural Research Service, U.S. Department of Agriculture, Watkinsville, GA, p. 29.

Bruce, R. R., L. A. Harper, R. A. Leonard, W. M. Snyder, and A. W. Thomas. 1975. A model for runoff of pesticides from small upland watersheds. *J. Environ. Qual.* 4(4):541–548.

Bufill, M. C. and M. J. Boyd. 1987. Predicting Runoff from Field Catchments Using a Parallel Reservoir Model. Seminar on Urban Drainage Modeling Data Base and Its Application, 22nd Congress. International Association for Hydraulic Research, Lousanne, pp. 65–70.

Burch, G. J., I. D. Moore, C. J. Barnes, J. M. Aveyard, and P. J. Barker. 1986. Modelling erosion hazard, a total catchment approach. In: Hydrology and Water Resource Symposium 1986. National Conference Publication No. 86/13. The Institution of Engineers, Brisbane, Australia. pp. 345–349.

Cahill, T. H., R. W. Pierson, Jr., and B. R. Cohen. 1979. Nonpoint source model calibration in Honey Creek watershed. EPA-600/3-79-054. U.S. Environmental Protection Agency, Athens, GA, p. 130.

Campbell, S. Y., C. W. Rose, and J.-Y. Parlange. 1984. Overland flow on converging and diverging surfaces. Proc. Conf. Australian Institution of Agricultural Engineers, National Conference Publication No. 84/6. The Institution of Engineers, Brisbane, Australia, pp. 299–303.

Capece, J. C., K. L. Campbell, and L. B. Baldwin. 1988. Estimating runoff peak rates from flat, high-water-table watersheds. *Trans. ASAE* 31(1):74–81.

Carroll, C., C. W. Rose, and S. J. Crawford. 1986. GUESS—Theory Manual. Division of Australian Environmental Studies, Griffith University, Brisbane, Australia, 4111.

Chery, D. L. 1963. Runoff studies with a physical model of a watershed. *Trans. Am. Geophys. Union* 44:869–870.

Chow, V. T. 1967. Laboratory studies of watershed hydrology. *Proc. Int. Hydrol. Symp.* 1:194–202.

Clark, C. O. 1945. Storage and the unit hydrograph. *Trans. ASCE* 110:1419–1488.

Clarke, R. T. 1973. A review of some mathematical models used in hydrology with observation on their calibration and use. *J. Hydrol.* 19:1–20.

Cleveland, J. G., B. W. Reid, and J. F. Harp. 1970. Evaluation of dispersed pollutional loads from urban areas. Bureau of Water Resources Research, Oklahoma University, Norman, OK, p. 213.

Connolly R. D., D. M. Silburn, C. A. A. Ciesiolica, and J. L. Foley. 1988. Simulating the hydrology of agricultural catchments using distributed parameter models. Paper presented at the Soil Management 88 Conference, September 18–21, Darling Downs Institute for Advanced Education, Toowoomba, Queensland.

Cook, D. J., W. T. Dickinson, and R. P. Rudra. 1986. GAMES: A model for watershed management of soil erosion and fluvial sedimentation. Microcomputer software demonstration. ASAE-NAR Annual Meeting. 1986, Laval University, Quebec, Canada.

Crawford, N. H. and A. S. Dunigian, Jr. 1973. Pesticide transport and runoff model for agricultural land. EPA-660/2-74-013. U.S. Environmental Protection Agency, Athens, GA, p. 211.

Crawford, N. H. and R. K. Linsley. 1966. Digital simulation in hydrology. Stanford watershed model IV. Department of Civil Engineering, Stanford University, Palo Alto, CA.

Crawford, S. J. 1986. GUESS—User Manual. Division of Australian Environmental Studies, Griffith University, Brisbane, Australia, 4111.

Crowder, B. M., H. B. Pionke, D. J. Epp, and C. E. Young. 1985. Using CREAMS and economic modelling to evaluate conservation practices: an application. *J. Environ. Qual.* 14(3):428–434.

Curtis, D. C. 1976. A deterministic urban storm water and sediment discharge model. In: Proc. of National Symposium on Urban Hydrology, Hydraulics and Sediment Control. University of Kentucky, Lexington, KY, pp. 151–152.

Davidson, J. M. and J. R. McDougal. 1973. Experimental and predicted movement of three herbicides in a water saturated soil. *J. Environ. Qual.* 2(4):428–433.

Davidson, J. M., R. S. Mansell, and D. R. Baker. 1973. Herbicides distributons within a soil profile and their dependence upon adsorption/desorption. *Proc. Soil Crop Sci. Soc. Fla.* 29.

Davidson, J. M., G. H Brusewiz, D. R. Baker, and A. L. Wood. 1975. Use of soil parameters for describing pesticide movement through soils. EPA-660/2-75-009. EPA Technological Service, U.S. Environmental Protection Agency, Athens, GA, p. 149.

Davidson, J. M., D. A. Graetz, P. S. C. Rao, and H. M. Selin. 1978a. Simulation of nitrogen movement, transformation and uptake in plant root zone. EPA-600/3-78-029. U.S. Environmental Protection Agency, Athens, GA.

Davidson, J. M., L. T. Ou, and P. S. C. Rao. 1978b. Adsorption, movement and biological degradation of high concentrations of selected pesticides in soils. In: Proc. 4th Annual Research Symposium, Land Disposal of Hazardous Wastes. EPA-600/9-78-016, pp. 233–244.

Dean, D. J. 1983. Potency Factor and Loading Functions for Predicting Agricultural Nonpoint Pollution in Agricultural Management and Water Quality. F. W. Schaller and G. W. Bailey (Eds.). Iowa State University Press, Ames.

DeCoursey D. G. 1982. ARS's Small Watershed Model. Paper No. 82-2094. American Society of Agricultural Engineers, St. Joseph, MI, p. 33.

DeCoursey, D. G. 1985. Mathematical models for non-point water pollution control. *J. Soil Water Conserv.* 40(5):408–413.

Dickinson, W. T. 1987. Water quality modeling — where are we? Paper presented at the AIC Conference, August 23–27, London, Ontario, Canada. Ontario Region of Canadian Society of Agricultural Engineers.

Dickinson, W. T., R. Pall, and G. J. Wall. 1984. GAMES—a model of identifying sources and amounts of soil erosion and fluvial sediment. In: Proc. Canadian Hydrology Symposium on Water Quality Evaluation within the Hydrological Cycle of Watersheds. Quebec, Canada, pp. 805–824.

Dickinson, W. T., R. P. Rudra, and G. J. Wall. 1985. Evaluation of alternative remedial measures strategies for controlling soil erosion and fluvial sedimentation. ASAE paper no. 85–2542, American Society of Agricultural Engineers, Chicago, IL.

Dickinson, W. T., R. P. Rudra, and D. J. Clark. 1986a. A delivery ratio for seasonal transport of sediment. In: Drainage Basin Sediment Delivery. International Association of Hydrological Sciences, Publ. No. 159, pp. 237–251.

Dickinson, W. T., R. P. Rudra, and G. J. Clark. 1986b. Identification of soil erosion and fluvial sediment problems. *Hydrol. Proc.* 1:111–124.

Dillaha, T. A. and D. B. Beasley. 1983. Distributed parameter modeling of sediment movement and particle size distribution. *Trans. ASAE* 26(6):1766–1772, 1777.

Diskin, M. H. 1967. A dispersion analogue model for watershed systems. *Proc. Int. Hydrol. Symp.* 1:38–45.

Donigian, A. S., Jr. 1982. Water quality modelling in relation to watershed hydrology. In: Modeling Component of Hydrologic Cycles. V. P. Singh (Ed.). Water Resources, Littleton, CO, pp. 343–382.

Donigian, A. S., Jr. 1986. Integration of runoff and receiving water models for comprehensive watershed simulation and analysis of agricultural management alternatives. In: Agricultural Nonpoint Source Pollution. A. Giorgina and F. Zingales (Eds.). Elsevier, New York, pp. 265–275.

Donigian, A. S., Jr. and N. H. Crawford. 1976a. Modelling nonpoint pollution from the land surface. EPA-600/3-76-083. EPA Technological Service, U.S. Environmental Protection Agency, Athens, GA.

Donigian, A. S., Jr. and N. H. Crawford. 1976b. Modelling pesticides and nutrients on agricultural lands. Report No. EPA-600/2-76-043. U.S. Environmental Protection Agency, Athens, GA.

Donigian, A. S., Jr. and N. H. Crawford. 1976c. Simulation of agricultural runoff. In: Proc. of the EPA Conference on Environmental Modeling and Simulation, Cincinnati, OH, 19–22, 1975. EPA/600/9-76/016. U.S. Environmental Protection Agency, pp. 151–156.

Donigian, A. S., Jr. and N. H. Crawford. 1977. Simulation of nutrient loading in surface runoff with the NPS model. EPA-660/3-77-065. U.S. Environmental Protection Agency, Athens, GA.

Donigian, A. S., Jr. et al. 1977. Agricultural Runoff Management Model (ARM), version II: refinement and testing. Report No. EPA-600/3-77-098. U.S. Environmental Protection Agency, Athens, GA.

Donigian, A. S., J. C. Imhoff, B. R. Bicknell, and J. L. Kittle. 1983. Application guide for hydrological simulation program FORTRAN (HSPF). Environmental Research Laboratory, U.S. Environmental Protection Agency, Athens, GA.

Enfield, C. G. and D. C. Shaw. 1975. Comparison of two predictive nonequilibrium one-dimensional models for phophorous sorption and movement through homogenous soils. *J. Environ. Qual. Anal.* 4(2):198–202.

Engman, E. T. 1974. Partial area hydrology and its application to water resources. *Water Resour. Res.* 10(3):464–473.

Farmer, W. J. and J. Letney. 1974. Volatilization losses of pesticides from soils. EPA-660/2-74-054. Office of Research and Development, U.S. Environmental Protection Agency, Athens, GA, p. 80.

Fisher, G. T., J. E. Ayers, H. N. Holtan, and D. L. Nelson. 1979. USDAHL-74 model as a planning tool. *Trans. ASAE* 22:1345–1352.

Fleming, G. and M. Fahmy. 1973. Some mathematical concepts for simulating the water and soil systems of natural watershed areas. Report No. H0-73-26. Department of Civil Engineering, Strathclyde University, Scotland.

Foster, G. R. 1976. Sedimentation. In: General Proc. of the Symposium on Hydraulics and Sediment Control. University of Kentucky, Lexington, KY.

Foster, G. R. 1980. Soil erosion modelling: special considerations for nutrient pollution evaluation of field sized area. In: Environmental Impact of Nonpoint Source Pollution. M. R. Overcash and J. M. Davidson (Eds.). Ann Arbor Science, Ann Arbor, MI, pp. 213–243.

Foster, G. R. 1982. Modelling the erosion process. In: Hydrologic Modelling of Small Watersheds. ASAE Monograph No. 5. Haan et al. (Eds.). American Society of Agricultural Engineers, St. Joseph, MI, pp. 297–379.

Foster, G. R. (Ed.). 1987. User Requirement: USDA. Water Erosion Prediction Project (WEPP). NSERL Report No. 1. National Soil Erosion Research Laboratory, Agricultural Research Service, U.S. Department of Agriculture, W. Lafayette, IN, p. 43.

Foster, G. R. and L. D. Meyer. 1972. Transport of soil particles by shallow flow. *Trans. ASAE* 15(1):99–102.

Foster, G. R. and L. D. Meyer. 1975. Mathematical simulation of upland erosion using fundamental erosion mechanics. In: Present and Prospective Technology for Predicting Sediment Yields and Sources. USDA-ARS-S-40. Agricultural Research Service, U.S. Department of Agriculture, pp. 190–207.

Foster, G. R., L. D. Meyer, and C. A. Onstad. 1977. A runoff erosivity factor and variable slope length experiments for soil loss estimates. *Trans. ASAE* 20:683–687.

Foster, G. R., L. J. Lane, J. D. Nowlin, J. M. Laflen, and R. A. Young. 1980. A model to estimate sediment yield from field-size areas: development of model. In: CREAMS: A Field-Scale Model for Chemicals, Runoff, and Erosion from Agricultural Management Systems. W. G. Knisel (Ed.). Conservation Research Report No. 26, SEA, U.S. Department of Agriculture, pp. 36–64.

Freeze, R. A. 1975. A stochastic-conceptual analysis of one-dimensional groundwater flow in nonuniform homogeneous media. *Water Resour. Res.* 11:725–741.

Freney, J. R. and A. S. Black. 1988. Importance of ammonia volatilization as a loss process. In: Advances in Nitrogen Cycling in Agricultural Ecosystems. J. R. Wilson (Ed.). Commonwealth Agricultural Bureau International, Wallingford, England, pp. 156–173.

Frere, M. H. 1975. Integrating chemical factors with water and sediment transport from a watershed. *J. Environ. Qual.* 4(1):12–17.

Frere, M. H. 1982. Modeling the quality of water from agricultural land. In: Hydrologic Modeling of Small Watersheds. ASAE Monograph No. 5. Haan et al. (Eds.). American Society of Agricultural Engineers, St. Joseph, MI, pp. 383–401.

Frere, M. H., C. A. Onstad, and H. N. Holtan. 1974. Modeling the movement of agriculture chemicals. In: Proc. 1974 Summer Computer Simulation Conference. pp. 271–274.

Frere, M. H., C. A. Onstad, and H. N. Holtan. 1975. ACTMO, an agricultural chemical transport model. ARS-H-3. Agricultural Research Service, U.S. Department of Agriculture, Washington, D.C., p. 54.

Freundlich. 1926.

Frink, C. R. 1969. Chemical and mineralogical characteristics of eutrophic lake sediment. *Soil Sci. Soc. Am. Proc.* 33:369–372.

Gburek, W. J. 1983. Hydrologic delineation of nutrient source contributing area. *J. Environ. Eng.* 109(5):1035–1047.

Green, W. H. and G. A. Ampt. 1911. Studies in soil physics. 1. The flow of air and water through soils. *J. Agric. Sci.* 4:1–24.

Hague, R., F. T. Lindstrom, V. H. Freed, and R. Sexton. 1968. Kinetic study of the sorption of 2-4D on some clays. *Environ. Sci. Technol.* 2:207–211.

Hairsine, P. B. and C. W. Rose. 1991. Rainfall detachment and deposition: sediment transport in the absence of flow-driven processes. *Soil Sci. Soc Am. J.* 55:320–324.

Haith, D. A. 1976. Land use and water quality in New York River. *J. Environ. Eng. Div. Am. Soc. Civ. Eng.* 102:1–15.

Haith, D. A. 1980. A mathematical model for estimating pesticide losses in runoff. *J. Environ. Qual.* 9(3):428–433.

Haith, D. A. 1986. Simulated regional variations in pesticide runoff. *J. Environ. Qual.* 15(1):5–8.

Haith, D. A. and J. V. Dougherty. 1976. Nonpoint source pollution from agricultural runoff. *J. Environ. Eng.* 102(EE5):1055–1069.

Haith, D. A. and R. C. Loehr. 1979. Effectiveness of soil and water conservation practices for pollution control. Report No. EPA-600/3-79-106. U.S. Environmental Protection Agency, Athens, GA.

Haith, D. A. and L. J. Tubbs. 1979. Modeling nitrogen export in rainfall and snowmelt runoff. In: Best Management Practices for Agriculture and Silviculture. R. C. Loehr, D. A. Haith, M. F. Walter, and C. S. Martin (Eds.). Ann Arbor Science, Ann Arbor, MI, pp. 665–685.

Haith, D. A. and L. J. Tubbs. 1981. Watershed loading functions for nonpoint sources. *J. Environ. Eng. Div. Am. Soc. Civ. Eng.* 107:121–137.

Hakonson, T. E., G. R. Foster, L. J. Lane, and J. W. Nyhan. 1986. The application of an agricultural water balance and erosion model in environmental science: a user perspective. In: Agricultural Nonpoint Source Pollution, Model Selection and Application. A. Glorgini and F. Zingales (Eds.). Elsevier, New York, pp. 173–190.

Hallam, P. M. 1976. Water quality studies on experimental catchments and an ecosystem. In: Hydrologic Techniques for Upstream Conservation. FAO Conservation Guide 2. Food and Agriculture Organization, Rome, Italy, pp. 73–87.

Hamaker, J. W. and J. M. Thompson. 1972. Adsorption. In: Organic Chemicals in Environment. C. A. I. Goring and J. W. Hamaker (Eds.). Marcel Dekker, New York, pp. 51–139.

Haque, R., F. T. Linstrom, V. H. Freed, and R. Sexton. 1968. Kinetic study of the sorption of 2-4D on some clays. *Environ. Sci. Technol.* 2:207–211.

Hearwole, C. D., K. L. Campbell, and A. B. Bottcher. 1987. Modified CREAMS hydrology model for costal plain flatwoods. *Trans. ASAE* 30(4): 1014–1022.

Hearwole, C. D., K. L. Campbell, and A. B. Bottcher. 1988. Modified CREAMS nutrient model for costal plain watersheds. *Trans. ASEA* 31(1):155–160.

Helyar, K. R. and D. N. Munns. 1975. Phosphorus flux in the soil plant system: a computer simulation. *Hilgardia* 43(4):103–130.

Higgins, R. W. 1976. The development of a general watershed-highway mathematical model. Master's thesis, University of South Florida, Tampa, FL.

Hirschi, M. C. and R. B. J. Barfield. 1988a. A physically based research erosion model. 1. Model development. *Trans. ASAE* 31(3):804–813.

Hirschi, M. C. and R. B. J. Barfield. 1988b. A physically based erosion model. II. Model sensitivity analysis and testing. *Trans. ASAE* 31(3):814–820.

Holtan, N. H. 1961. A concept for nitrification estimates in watershed engineering. Report No. ARS-41-51. Agricultural Research Service, U.S. Department of Agriculture, Washington, D.C.

Holtan, H. N. 1965. A model for computing watershed retention from soil parameters. *J. Soil Water Conserv.* 20(3):91–94.

Holtan, H. N. 1979. Procedures manual for sediment, phosphorous and nitrogen transport. Computation with USDAHL. Report No. MP 943. Maryland Agricultural Experimental Station, University of Maryland, College Park, MD.

Holtan, H. N. and N. C. Lopez. 1971. USDAHL-70 model for watershed hydrology. Technical Bulletin No. 1435. Agricultural Research Service, U.S. Department of Agriculture, Washington, D.C.

Holtan, H. N., G. J. Stiltner, W. H. Henson, and N. C. Lopez. 1975. USDAHL-74: revised mode of watershed hydrology. Technical Bulletin No. 1518. Agricultural Research Service, U.S. Department of Agriculture, Washington, D.C.

Holtan, H. N., J. P. Ormsby, and G. T. Fisher. 1977. Application of a Maryland version of USDAHL-74 to a watershed in Prince Geoge's County, MD. In: Proc. Watershed Research in Eastern North America. A workshop sponsored by the Smithsonian.

Horton, S. E., 1939. Estimating soil permeability rates. *Proc. Soil Sci. Soc. Am.* 5:399–417.

Huber, W. C., J. P. Heaney, and S. J. Nix. 1981. Stormwater management model. User's manual version III. Department of Environmental Engineering and Science, University of Florida, FL.

Huck, P. M. and G. J. Farquhar. 1974. Water quality models using the Box-Jenkins method. *J. Environ. Eng.* 100:733–752.

Huff, D. D., R. V. Luxmore, V. B. Mankin, and C. L. Begovich. 1977. TEHM. A terrestrial ecosystem hydrology model ORNL/NSF/EATC-17. Oak Ridge National Laboratory, Oak Ridge, TN, p. 152.

Huggins, L. F. and J. Monke. 1966. The mathematical simulation of the hydrology of small watersheds. Technical Report 82. Water Resources Research Center, Purdue University, West Lafayette, IN, p. 59.

Hydrocomp, Inc. 1969. Hydrocomp Simulation Programming (HSP): Operation Manual, 2nd ed. Hydrocomp, Inc., Palo Alto, CA, pp. 1–1 to 1–27 and 3–5 to 3–16.

Hydrocomp, Inc. 1975. Hydrocomp Simulation Program Operation Manual, 4th ed. Hydrocomp, Inc., Palo Alto, CA, p. 115.

Hydrocomp, Inc. 1977. Hydrocomp Water Quality Operation Manual. Hydrocomp, Inc., Palo Alto, CA.

Ishi-hara, T. 1961. Runoff analysis by analogue computer. In: General Assembly of International Association for Hydraulic Research, 9th, Dubrovnik, 1961, pp. 749–759.

Johnson, M. G. et al. 1978. Management information base and overview modelling. PLUARG Report. International Joint Commission U.S.-Canada, Windsor, Ontario, Canada.

Kang, F. F., L. T. Fan, E. S. Lee, and L. E. Erickson. 1970. Modelling feedlot runoff pollution. 1. Analogue simulation. *Trans. ASAE* 13:859–863.

Kenimer, A. L., S. Mostaghimi, T. A. Dillaha, and V. O. Shanholtz. 1989. PLIERS: Pesticide Losses in Erosion and Runoff Simulator. *Trans. ASAE* 32:127–136.

Khaleel, R. 1980. Hydrologic processes affecting nonpoint pollution from agricultural lands. In: Environment Impact of Nonpoint Source Pollution. M. R. Overcash and J. M. Davidson (Eds.). Ann Arbor Science, Ann Arbor, MI, pp. 375–396.

Khaleel, R., G. R. Foster, K. R. Reddy, M. R. Overcash, and P. W. Westerman. 1979a. A nonpoint source model for land areas receiving animal wastes. III. A conceptual model for sediment and manure transport. *Trans. ASAE* 22:1353–1360.

Khaleel, R., G. R. Foster, K. R. Reddy, M. R. Overcash, and P. W. Westerman. 1979b. A nonpoint source model for land areas receiving animal wastes. IV. Model inputs and verification for sediment and manure transport. *Trans. ASAE* 22:1362–1368.

Knisel, W. G. (Ed.). 1980. CREAMS: A field-scale model for chemicals, runoff and erosion from agricultural management system. Report No. 26. USDA.

Knisel, W. G., D. C. Moffitt, and T. A. Dumper. 1985. Representing seasonally frozen soil with CREAMS model. *Trans. ASAE* 28(5):1487–1493.

Kumar, S. and R. A. Rastogi. 1987. A conceptual catchment model for estimating suspended sediment flow. *J. Hydrol.* 95:155–163.

Lane, L. J. 1982a. Development of a procedure to estimate runoff and sediment transport in ephemeral streams. In: Recent Developments in the Explanation and Prediction of Erosion and Sediment Yield. Proc. Exeter Symp. IASH Publ. No. 137, pp 275–282.

Lane, L. J. 1982b. Distributed model for small semi-arid watersheds. *J. Hydraul. Div. Am. Soc. Civ. Eng.* 108(HY10):1, 114–1, 131.

Lane, L. J., J. E. Gilley, M. Nearing, and A. D. Nicks. 1988. The USDA Water Erosion Prediction Project. Paper presented at the American Society of Civil Engineers Hydraulic Engineering Conference, August.

Lane, L. J. and M. A. Nearing. 1989. USDA–Water Erosion Prediction Project: Hillslope profile model documentation. NSERL Report No. 2. USDA-ARS National Soil Erosion Research Lab, West Lafayette, IN.

Lane, W. L. 1975. Extraction of Information on Inorganic Water Quality. Hydrology paper 73. Colorado State University, CO.

Laurensen, E. M. and R. G. Mein. 1985. RORB Version 3 Runoff Program User Manual, 2nd ed. (revised March 1985). Department of Civil Engineering, Monash University, Calyton, Victoria, Australia.

Laurensen, E. M., C. P. Codner, and R. G. Mein. 1985. Giralang/Gungahlin paired catchment study. A review. Department of Housing and Construction. ACT, Australia.

Leavsley, G. H., R. W. Lichty, B. M. Troutman, and L. G. Saindon. 1983. Precipitation-runoff modeling system: User's manual. Water Resources Investigations Report 83-4238. U.S. Geological Survey, Washington, D.C., p. 207.

Leonard, R. A. and W. G. Knisel. 1986. Selection and application of models for nonpoint source pollution and resource conservation. In: Agricultural Nonpoint Source Pollution: Model Selection and Application. A. Giorgini and F. Zingales (Eds.) Dev. Environ. Model. 10, Elsevier, pp. 213–229.

Leonard, R. A. and R. D. Wauchope. 1980. The pesticide submodel. In: CREAMS: A Field-Scale Model for Chemicals, Runoff, and Erosion from Agricultural Management Systems. Knisel, W. G. (Ed.). Conservation Research Report No. 26. SEA, U.S. Department of Agriculture, Washington, D.C., pp. 88–112.

Leonard, R. A., G. W. Bailey, and R. R. Swank, Jr. 1976. Transport, detoxification, fate and effects of pesticides in soil and water environments. In: Land Application of Waste Materials. Soil Conservation Soc. Am. Ankeny, IA, p. 48–78.

Leonard, R. A., G. W. Longdale, and W. G. Flemming. 1979. Herbicide runoff from upland piedmont watersheds — data and implication for modelling pesticide transport. *J. Environ. Qual.* 8:223–229.

Leonard, R. A., W. G. Knisel, and D. A. Sull. 1987. GLEAMS: groundwater loading effects of agricultural management systems. *Trans. ASAE* 30(5):1403–1418.

Leytham, K. M. and R. C. Johanson. 1979. Watershed Erosion and Sediment Transport (WEST) Model. Environmental Research Laboratory, U.S. Environmental Protection Agency, Athens, GA.

Li, R. M. 1974. Mathematical modelling of response from small watershed. Ph.D. dissertation. Department of Civil Engineering, Colorado State University, Fort Collins, CO.

Li, R. M., R. K. Simons, and L. Y. Shio. 1977. Mathematical modelling of on-site soil erosion. In: Proc. of International Symposium on Urban Hydrology, Hydraulics and Sediment Control. UKYBU114. University of Kentucky, Lexington, KY, pp. 87–94.

Lighthill, M. J. and G. B. Whitham. 1955. Kinematic waves. I. *Proc. R. Soc. London A:* 229:281–316.

Linden D. R. 1979. A model to predict soil water storage as affected by tillage practices. Ph.D. dissertation, Department of Soil Science, University of Minnesota, St. Paul.

Lindstrom, F. T. and L. Boersma. 1973. A theory on the mass transport of previously distributed chemicals in water saturated, sorbing porous media. 111. Exact solution for first order kinetic sorption. *Soil Sci.* 115(1):5–10.

Lindstrom, F. T., R. Hague, and W. R. Coshow. 1970. Adsorption from solution. 111. A new model for the kinetics of adsorption-desorption processes. *J. Phys. Chem.* 74: 495–502.

Littleboy, M., D. M. Silburn, D. M. Freebairn, D. R. Woodruff, and G. L. Hammer. 1989. PERFECT: A Computer Simulation Model of Productivity Erosion Runoff Function to Evaluate Conservation Techniques. Bulletin QB89005. Queensland Department of Primary Industries, Brisbane, Australia.

Loch, R. J., D. M. Silburn, and D. M. Freebairn. 1989. Evaluation of the CREAMS model. II. Use of rainulator data to derive soil erodibility parameters and prediction of field scale soil losses using derived parameters. *Aust. J. Soil Res.* 27:563–576.

Logan, T. J. 1980. The role of soil and sediment chemistry in modelling nonpoint source of phosphorous. In: Environmental Impact of Nonpoint Source Pollution. M. R. Overcash and J. M. Davidson (Eds.). Ann Arbor Science, Ann Arbor, MI, pp. 189–210.

Logan, T. J. and R. Stiefel. 1979. Maumee River Basin Pilot Watershed Study. PLUARG Task C. Final Report. U.S. Environmental Protection Agency, Region V, Chicago.

Lorber, M. N. and L. A. Mulkey. 1982. An evaluation of three pesticide runoff loading models. *J. Environ. Qual.* 11(3):519–529.

Lovell, C. J. and C. W. Rose. 1986. Measurement of the settling velocities of soil aggregates using a modified bottom withdrawal tube, AES Working Paper 4/86, Griffith University, Nathan, Brisbane.

Lovell, C. J. and C. W. Rose. 1988. Measurement of soil aggregate settling velocities I. A modified Bottom Withdrawal tube method. *Aust. J. Soil Res.* 26:55–71.

Mazurak, A. P., L. Chesnin, and A. E. Tiarks. 1975. Detachment of soil aggregates by simulated rainfall from heavily manured soils in eastern Nebraska. *Soil Sci. Soc. Am. Proc.* 39(4):732–736.

McElroy, A. D., S. Y. Chiu, J. W. Nebgen, A. Aleti, and F. W. Bennett. 1976. Loading functions for assessment of water pollution from nonpoint sources. EPA 600/2-76-151. U.S. Environmental Protection Agency, Washington, D.C.

Mehran, M. and K. K. Tanji. 1974. Computer modelling of nitrogen transformations in soils. *J. Environ. Qual.* 3(4):391–399.

Mein, R. G. and C. L. Larson. 1971. Modeling the infiltration components of the rainfall-runoff process. Graduate School Bulletin 43. WRRC, University of Minnesota.

Menzel, R. G. 1980. Enrichment ratios for water quality modeling. In: CREAMS: A Field-Scale Model for Chemicals, Runoff, and Erosion from Agricultural Management Systems. W. G. Knisel (Ed.). Conservation Research Report No. 26. SEA, U.S. Department of Agriculture, Washington, D.C.

Metcalf and Eddy, 1971. University of Florida and Water Resource Engineering, Inc., Stormwater Management Model (4 volumes). U.S. Environmental Protection Agency, Washington, D.C.

Meyer, L. D. and W. H. Wischmeier. 1969. Mathematical simulation of the process of soil erosion by water. *Trans. ASAE* 12(6):754–762.

Misra, R. K. and C. W. Rose. 1989. Manual for Use of Program GUEST, Version 2.7. Australian Environmental Studies, Griffith University, Brisbane, Australia, p. 62.

Mockus, V. 1964. SCS National Engineering Handbook. 4. Hydrology. U.S. Soil Conservation Service, Washington, D.C., p. 548.

Mockus, V. 1972. Estimation of direct runoff from storm rainfall. In: SCS National Engineering Handbook. U.S. Soil Conservation Service, Washington, D.C., pp. 10.1–10.24.

Moore, I. D. 1981. Infiltration equations modified for surface effects. *J. Irrig. Drain. Div. Am. Soc. Civ. Eng.* 107(IR1):71–86.

Moore, I. D. 1985. Kinematic overland flow: generalization of Rose's approximate solution. *J. Hydrol.* 82:233–245.

Moore, I. D. and G. J. Burch. 1986. Sediment transport capacity of sheet and rill flow: application of unit stream power theory. *Water Resour. Res.* 22:1350–1360.

Moore, I. D. and J. D. Eigel. 1981. Infiltration into two layered soil profiles. *Trans. ASAE* 24(6):1496–1503.

Moore, I. D. and P. I. A. Kinnell. 1987. Kinematic overland flow — generalisation of Rose's approximate solution, Part II. *J. Hydrol.* 92:251–362.

Moore, I. D. and C. L. Larson. 1980. An infiltration model for cultivated soils. *Trans. ASAE* 23:1460–1467.

Moore, I. D., S. M. Mackay, P. J. Wallbrink, G. H. Burch, and E. M. O'Loughlin. 1986. Hydrologic characteristics and modelling of a small forested catchment in southeastern New South Wales. Pre-logging condition. *J. Hydrol.* 83:307–335.

Morgan, R. P. C. and H. J. Finney. 1982. Stability of agricultural ecosystems: validation of a simple model for soil erosion assessment. Int. Inst. Applied Systems Analysis, Collaborative Paper No. CP-82-76.

Morgan, R. P. C., D. D. V. Morgan, and H. J. Finney. 1984. A predictive model for the assessment of soil erosion risk. *J. Agric. Eng. Res.* 30:245–253.

Mulkey, L. A., R. F. Carsel, and C. N. Smith. 1986. Development, testing and applications of nonpoint source models for evaluation of pesticides risk to the environment. In: Agricultural Nonpoint Source Pollution, Model Selection and Application. A. Giorgini and F. Zingales (Eds.) Dev. Environ. Model. 10, 383–398.

Mulkey, L. A. and J. W. Falco. 1977. Sedimentation and erosion control implications for water quality management. In: Proc. National Symposium on Soil Erosion and Sedimentation. American Society of Civil Engineers, Chicago, IL.

Nash, J. E. 1957. The form of the instantaneous unit hydrograph. *Bull. Int. Assoc. Sci. Hydrol.* 111:114–121.

Negev, M. 1967. A sediment model on a digital computer. Technical Report No. 62. Department of Civil Engineering, Stanford University, Palo Alto, CA.

Nezafati, H., D. S. Bowles, and L. D. James. 1986a. Hydro salinity modelling of suspended sediments. *J. Environ. Eng.* 112(4):663–676.

Nezafati, H., D. S. Bowles, and J. P. Riley. 1986b. Salt release model for suspended sediments. *J. Environ. Eng. Div. Am. Soc. Civ. Eng.* 112(4):647–662.

Novotny, V. 1986. A review of hydrologic and water quality models used for simulating agricultural pollution. In: Agricultural Nonpoint Source Pollution, Model Selection and Application. A. Giorgini and F. Zingales (Eds.) Dev. Environ. Model. 10, Elsevier, pp.9–35.

Novotny, V. and G. Chesters. 1981. Handbook of Nonpoint Pollution: Sources and Management. Van Nostrand Reinhold, New York.

Novotny, V., H. Tran, G. V. Simsiman, and G. Chesters. 1978. Mathematical modelling of pond runoff contamination by phosphorus. *J. Water Pollut. Control Fed.* 50: 101–112.

Nutter, W. L., T. Tkacs, P. B. Bush, and D. G. Neary. 1984. Simulation of herbicide concentrations in stormflow from forested watersheds. *Water Res. Bull.* 20(6):851–857.

Nyhan, J. W. and L. J. Lane. 1982. Use of a state of the art model in generic designs of shallow land repositories for low-level wastes. Waste Management 82. R. G. Post (Ed.). University of Arizona Press, Tucson, AZ, pp. 235–244.

O'Conner, D. J., R. V. Thonnan, and D. N. Ditoro. 1973. Dynamic water quality forecasting and management. EPA-660/3-73-009. EPA Technological Service, U.S. Environmental Protection Agency, Washington, D.C.

Oddson, J. K., J. Letey, and L. V. Weeks. 1970. Predicted distribution of organic chemicals in solution and adsorbed as a function of position and time for various chemical and soil properties. *Soil Sci. Soc. Am. Proc.* 34(3):412–417.

O'Loughlin, E. M. 1981. Saturation regions in catchments and their relations to soil and topographic features. *J. Hydrol.* 53:229–246.

O'Loughlin, E. M. 1986. Prediction of surface saturation zones in natural catchments by topographic analyses. *Water Resour. Res.* 22:794–804.

Onstad, C. A. and G. R. Foster. 1975. Erosion modelling on a watershed. *Trans. ASAE* 18:288–292.

Osman, A. 1987. Pollutant washoff by overland flow. *J. Environ. Eng.* 113(4):811–823.

Parlange, J.-Y., C. W. Rose, and G. Sander. 1981. Kinematic flow approximation of runoff on a plane: an exact analytic solution. *J. Hydrol.* 52:171–176.

Pathak, C. S., F. R. Crow, and R. L. Bengtson. 1984. Comparative performance of two runoff models on a grassland watershed. *Trans. ASAE* 28:397–402, 406.

Patrick, W. J. and I. C. Mahapatra. 1968. Transformation and availability of nitrogen and phosphorus in waterlogged soils. *Adv. Agron.* 20:323–359.

Patterson, M. R., C. L. Beguvich, and D. R. Jackson. 1976. Environmental transport modelling of pollutants in water and soil. In: Symposium on Nonbiological Transport and Transformation of Pollutant on Land and Water. National Bureau of Standards, Gaithersburg, MD.

Perroux, K. M., D. E. Smiles, and I. White. 1981. Water movement in uniform soils during constant-flux infiltration. *Soil Sci. Soc. Am. J.* 45:237–240.

Philip, J. R. 1954. An infiltration equation with physical significance. *Soil Sci.* 77:153–157.

Porter, K. S. 1975. Nitrogen and phosphorus food production waste and environment. Ann Arbor Science, Ann Arbor, MI, p. 372.

Proffitt, A. P. B. 1988. The influence of rainfall detachment, runoff entrainment and sediment deposition processes on sediment concentration and settling velocity characteristics of eroded soil. Ph.D. thesis. Griffith University, Brisbane, Australia.

Proffitt, A, P. B., C. W. Rose, and P. B. Hairsine. 1991. Rainfall detachment and deposition: experiments with low slope and significant water depths. *Soil Sci. Soc.Am. J.* 55:325–332.

Rallison, R. E. 1980. Origin and evolution of the SCS runoff equation. In: Symposium on Watershed Management 1980, Vol. II. American Society of Civil Engineers, pp. 912–924.

Rao, P. S. C. and J. M. Davidson. 1979. Adsorption and movement of selected pesticides at high concentrations in soils. *Water Res.* 13:375–380.

Rao, P. S. C. and J. M. Davidson. 1980. Estimation of pesticide retention and transformation parameters required in nonpoint source pollution models. In: Environmental Impact of Nonpoint Source Pollution. M. R. Overcash and J. M. Davidson (Eds.). Ann Arbor Science, Ann Arbor, MI, p. 23–67.

Rao, P. S. C. and J. M. Davidson. 1982. Retention and transformation of selected pesticides and phosphorus in soil water systems. A critical review. EPA 600/3-82-060. U.S. Environmental Protection Agency, Athens, GA.

Rao, P. S. C., J. M. Davidson, R. E. Jessup, and H. M. Selim. 1979. Evaluation of conceptual models for describing non-equilibrium adsorption-desorption of pesticides during steady flow in soils. *Soil Sci. Soc. J.* 43:22–28.

Reddy, K. R., R. Khaleel, M. R. Overcash, and P. W. Westerman. 1979a. A nonpoint source model for land areas receiving animal waste. I. Mineralization of organic N. *Trans. ASAE* 22(4):863–872.

Reddy, K. R., R. Khaleel, M. R. Overcash, and P. W. Westerman. 1979b. A nonpoint source model for land areas receiving animal waste. II. Ammonia volatilization. *Trans. ASAE* 22:1398–1405.

Renard, K. G. and E. M. Laursen. 1975. Dynamic behavior model of ephemeral stream. *J. Hydraul. Div. Proc. Am. Soc. Civ. Eng.* 101(HY5):511–528.

Ritchie, J. T. 1972. A model for predicting evaporation from a row crop with incomplete cover. *Water Resour. Res.* 8:1204–1213.

Roesner, L. A., P. A. Giguerre, and D. E. Evensum. 1977. Computer Program Documentation for Stream Quality Model QUAL II. Water Resource Engineering, Walnut Creek, CA.

Rogowski, A. S. and T. Tamura. 1970. Erosional behaviour of cesium-137. *Health Phys.* 18:467–477.

Rose, C. W. 1984. Recent advances in research soil erosion processes. In: Proc. National Soil Conference, Brisbane, Australia. pp. 2–26.

Rose, C. W. 1985. Developments in soil erosion and deposition models. *Adv. Soil Sci.* II:1–63.

Rose, C. W. and R. C. Dalal. 1988. Erosion and runoff of nitrogen. In: Advances in Nitrogen Cycling in Agricultural Ecosystems. J. R. Wilson (Ed.). Commonwealth Agricultural Bureau International, Wallingford, England.

Rose, C. W. and P. B. Hairsine. 1988. Processes of water erosion. In: Flow and Transport in the Natural Environment. W. L. Steffen, O. T. Denmead, and I. White (Eds). Springer-Verlag, New York, pp. 312–326.

Rose, C. W., F. W. Chichester, J. R. Williams, and J. T. Ritchie. 1982a. A contribution to simplified models of field solute transport. *J. Environ. Qual.* 11:146–150.

Rose, C. W., F. W. Chichester, J. R. Williams, and J. T. Ritchie. 1982b. Application of an approximate analytic method of computing solute profiles with dispersion in soils. *J. Environ. Qual.* 11:151–155.

Rose, C. W., P. B. Hairsine, A. P. B. Proffitt, and R. K. Misra. 1990. Interpreting the role of soil strength in erosion processes. *Catena Suppl.* 17:155–165.

Rose, C. W., J. R. Williams, G. C. Sander, and D. A. Barry. 1983a. A mathematical model of soil erosion and deposition processes. I. Theory for a plane land element. *Soil Sci. Soc. Am. J.* 47:991–995.

Rose, C. W., J.-Y. Parlange, G. C. Sander, S. Y. Campbell, and D. A. Barry. 1983b. A kinematic flow approximation to runoff on a plane: an approximate analytic solution. *J. Hydrol.* 62:363–369.

Rosenthal, K. M., B. J. White, and R. D. Berndt. 1976. A versatile computer model for simulating the soil water balance of cropping systems. Division of Land Utilisation Technical Report 17. Queensland Department of Primary Industries, Brisbane, Australia.

Rosewell, C. J. and K. Edwards. 1988. SOILOSS: A program to assist in the selection of management practices to reduce erosion. Technical Handbook No. 11. Soil Conservation Service of New South Wales, Australia.

Ross, B. B. et al. 1978. A model for evaluating the effect of land uses on flood flow. Bulletin 85. Virginia Water Resources Research Center. Virginia Polytechnic Institute and State University, Blacksburg, VA.

Rousseau, A., W. T. Dickinson, and R. P. Rudra. 1985. A phosphorus transport model for small agricultural watersheds. ASAE Paper No. 85-2526. American Society of Agricultural Engineers, Chicago, IL.

Rudra, R. P., W. T. Dickinson, and G. J. Wall. 1985. Application of the CREAMS model in southern Ontario conditions. *Trans. ASAE* 28(4):1233–1240.

Saxton, K. E., P. F. Brooks, R. Richmond, and J. S. Romberger. 1984. User manual for SPAW, A Soil-Plant-Air-Water model. SEA-AR, U.S. Department of Agriculture, Washington, D.C.

Shaffer, M. J. 1985. Simulation model for soil erosion-productivity relationships. J. Environ. Qual. 14(1):144–150.

Shaffer, M. J. and W. E. Larson (Eds.) 1982. Nitrogen-tillage-residue management (NTRM) model. Technical documentation. Res. Rep. USDA-ARS, St. Paul, MN.

Shaffer, M. J. amd F. J. Pierce. 1982. Nitrogen-tillage-residue management (NTRM) model: user's manual Res. Rep. USDA-ARS, St. Paul, MN

Shaffer, N. J., S. C. Gupta, D. R. Linden, J. A. E. Molina, C. E. Clapp, and W. E. Larson. 1983. Nitrogen-Tillage-Residue Management model (NTRM). In: Analysis of Ecological Systems. State of the Art in Ecological Methods. W. K. Lauenroth G. V. Skogerboe, and M. Flug (Eds.). Elsevier, New York, pp. 525–544.

Shah, D. B., G. A. Coulman, L. T. Novak, and B. G. Ellis. 1975. A mathematical model for phosphorus movement in soils. *J. Environ. Qual.* 4(1):87–92.

Sharpley, A. N. 1980. The enrichment of soil phosphorus in runoff sediments. *J. Environ. Qual.* 9(3):521–526.

Sharpley, A. N. 1983. Effect of soil properties on the kinetics of P desorption. *Soil Sci. Soc. Am. J.* 47:462–467.

Sharpley, A. N., L. R. Ahuja, M. Yamamoto, and R. G. Menzel. 1981a. The kinetics of phosphorus desorption from soil. *Soil Sci. Soc. Am. J.* 45:493–496.

Sharpley, A. N., L. R. Ahuja, and R. G. Menzel. 1981b. The release of soil phosphorus to runoff in relation to the kinetics of desorption. *J. Environ. Qual.* 10:386–391.

Sharpley, A. N., S. J. Smith, W. A. Berg, and J. R. Williams. 1985. Nutrient runoff losses as predicted by annual and monthly soil sampling. *J. Environ. Qual.* 14(3):354–360.

Shubinski, R. P. and L. A. Roesner. 1983. Stormwater models. In: Stormwater Management in Urbanizing Area. Prentice-Hall, Englewood Cliffs, NJ.

Silburn, D. M. and R. J. Loch. 1990. Present capabilities and constraints in modelling soil erosion. In: Proc. of the 5th Australian Soil Conservation Conference, Perth, Australia, pp. 187–199.

Silburn, D. M., R. J. Loch, R. D. Connelly, and G. D. Smith. 1990. Erosional stability of waste rock dumps at the proposed Coronation Hill Mine. Consultant's Report to Resource Assessment Commission. Australian Government Publication Series, Canberra, Australia.

Simons, D. B. and R. M. Li. 1980. Modelling of sediment nonpoint source pollution from watersheds. In: Environmental Impact of Nonpoint Source Pollution. M. R. Overcash and J. M. Davidson (Eds.). Ann Arbor Science, Ann Arbor, MI, pp. 341–374.

Simons, D. B., R. M. Li, and M. A. Stevens. 1975. Development of models for predicting water and sediment routing and yield from storms on small watersheds. CER7475 DBS-RML-MAS24. Civil Engineering Department, Colorado State University, CO.

Simons, D. B., R. M. Li, and K. G, Eggert. 1977. Stormwater and sediment runoff simulation for upland watersheds using analytical routing technique. Vol. 1: Water routing and yield. Colorado State University Rep. CER77-78DBS-RML-KGE16.

Smith, C. N. 1980. Sediment-pollutant relationships in runoff from selected agricultural, suburban and urban watersheds. EPA-600/3-80-022. U.S. Environmental Protection Agency, Athens, GA.

Smith C. N., R. A. Leonard, G. W. Langdale, and G. W. Baily. 1978. Transport of agricultural chemicals from small upland piedmont watersheds. EPA-600/3-78-056. U.S. Environmental Protection Agency, Athens, GA, p. 345.

Smith, R. E. and R. H. B. Hebbert. 1983. Mathematical simulation of interdependent surface and subsurface hydrologic processes. *Water Resour. Res.* 19(4):987–1001.

Smith, S. J., L. B. Young, and G. E. Miller. 1976. Field evaluation of soil nitrogen mineralization potential. *Soil Sci. Soc. Am. Proc.* 40:

Sonzogni, W. C. et al. 1980. WATERSHED: a management technique for choosing among point and nonpoint control strategies. In: Proc. of the Seminar on Water Quality Management. U.S. Environmental Protection Agency, Chicago, IL.

South Florida Water Management District (SFWMD). 1983. Permitting information manual. IV. Management and storage of surface waters. South Florida Water Management District, West Palm Beach, FL.

Southerland, E., J. P. Hartigan, and C. W. Randall. 1981. A continuous simulation modelling approach to nonpoint pollution management. Proc. Industrial Wastes and Research Symp., 54th WPLF Conference, Detroit, MI, October, 1981.

Springer, E. P., C. W. Johnson, K. R. Coorley, and D. C. Robertson. 1984. Testing of the SPUR hydrology component on rangeland watersheds in southwest Idaho. *Trans. ASAE* 27:1040–1054.

Steele, T. D. 1976. A bivariate-regression model for estimating chemical composition of streamflow of groundwater. *Hydrol. Sci. Bull.* 21:149–161.

Stephens, T. C. and W. C. Mills. 1965. Using the Cypress Creek formula to estimate runoff rates in the southern coastal plains and adjacent flatwoods land resource area. ARS Publication No. 41-95. Agricultural Research Service, U.S. Department of Agriculture, Washington, D.C.

Stoltenberg, N. K. and J. L. White. 1953. Selective loss of plant nutrients by erosion. *Soil Sci. Soc. Am. Proc.* 17:406–410.

Storm, D. E., T. A. Dillaha, III, S. Mostaghimi, and V. U. Shanholtz. 1988. Modelling phosphorus transport in surface runoff. *Trans. ASAE* 31(1):117–127.

Sweeney, D. W., A. B. Bottcher, K. L. Campbell, and D. A. Graetz. 1985. Measured and CREAM-predicted nitrogen losses from tomato and corn management systems. *Water Resour. Bull.* 21(5):867–873.

Syers, J. K., R. Shah, and T. W. Walker. 1969. Phosphorus fractionation in two alluvial soils and particle size separates. *Soil Sci.* 108:283–289.

Syers, J. K., R. F. Harris, and D. E. Armstrong. 1973. Phosphate chemistry in lake sediments. *J. Environ. Qual.* 2:1–14.

Tholin, A. L. and C. J. Keifer. 1960. Hydrology of urban runoff. *Trans. Am. Soc. Civ. Eng.* 125:1308–1355.

Thomas, J. M. and W. J. Thomas. 1967. Introduction to the Principles of Heterogeneous Catalysis. Academic Press, New York.

Todini, E. and J. R. Wallis. 1977. Using CLS for daily or longer period rainfall-runoff modelling. In: Mathematical Models for Surface Water Hydrology. T. A. Ciriani et al. (Eds.). John Wiley & Sons, New York, pp. 149–168.

Townley, L. R. and K. J. Bligh. 1990. Evaluation of physically-based rainfall-runoff models using field data from Berkshire Valley, Western Australia. In: Proc. of the 5th Australian Soil Conservation Conference, Perth, Australia.

True, H. A. 1972. Planning models for non-point runoff assessment. In: Proc. of the Conference on Environmental Modeling and Simulation. W. R. Ott (Ed.). EPA 600/9-76-016. U.S. Environmental Protection Agency, Cincinatti, OH, pp. 74–76.

True, H. A. 1976. Planning models for nonpoint runoff assessment. In:Proc. Environmental Modelling and Simulation. W. R. Ott (Ed.) U.S. Environmental Protection Agency, Rep. No. EPA/600/9-76-016, Cincinnati, OH.

U.S. Army Corps of Engineers. 1977. Storage, Treatment, Overflow, Runoff Model. "STORM" User's Manual. Hydrology Engineering Center, Davis, CA, p. 170.

U.S. Army Corps of Engineers. 1982. CE-Qual-R7: A Numerical One Dimensional Model of Reservoir Water Quality. User's Manual. Waterways Experimental Station Environmental Laboratory, Vicksburg, MS, p. 528.

U.S. Department of Agriculture, Agricultural Research Service. 1986. AGNPS-PC. Agricultural Nonpoint Source Pollution Program for Personal Microcomputer, a User Guide. Developed for the Minnesota Pollution Control Agency by the Department of Soil Science, University of Minnesota. Agricultural Research Service, U.S. Department of Agriculture, Morris, MN, and Soil Conservation Service, p. 37.

U.S. Department of Agriculture, Soil Conservation Service. 1976. Hydrology Guide for Minnesota. St. Paul, MN, p. 165.

U.S. Department of Agriculture, Soil Conservation Service. 1980. Interim runoff procedure for Florida. Fluoride Bulletin 210-1-2. Soil Conservation Service, U.S. Department of Agriculture, Gainesville, FL.

U.S. Department of Commerce. 1976. Proc. of EPA Conference on Environmental Modelling and Simulation, Cincinnati, OH, April 19–22, 1975. PB-257142. Booz-Allen Applied Research, Inc., Chicago, IL.

U.S. Department of Commerce. 1976b. Modelling pesticides and nutrients on agricultural lands. PB-250566. Hydrocomp, Inc., Environmental Research Laboratory, Athens, GA.

U.S. Department of Commerce. 1976c. Modelling nonpoint pollution from the land surface. Hydrocomp, Inc., Environmental Research Laboratory, Athens, GA.

U.S. Salinity Laboratory Staff. 1954. Diagnosis and improvement of saline and alkaline soils. Agricultural Handbook 60 (reprinted 1969). U.S. Department of Agriculture, Washington, D.C.

Van de Griend, A. A. and E. T. Engman. 1985. Partial area hydrology and remote sensing. *J. Hydrol.* 81:211–251.

Van Genuchten, M. T., J. M. Davidson, and P. J. Wieringe. 1974. An evaluation of kinetic and equilibrium equations for the prediction of pesticide movement through porous media. *Soil Sci. Soc. Am. J.* 38:29–35.

Vertessy, R. A., C. J. Wilson, E. M. O'Loughlin, W. R. Dawes, and D. L. Short. 1990. Development in digital terrain and hydrologic modelling techniques and their potential for managing soil erosion and dryland salinity. In: Proc. of the 5th Australian Soil Conservation Conference. Western Australian Department of Agriculture, Perth, Australia.

Wall, G. J., W. T. Dickinson, and L. J. P. Van Valiet. 1982. Agriculture and water quality in Canadian Great Lakes Basin. II. Fluvial sediments. *J. Environ. Qual.* 11(3):482–487.

Weber, J. B., P. J. Shea, and H. J. Strek. 1980. An evaluation on nonpoint sources of pesticides pollution in runoff. In: Environmental Impact of Nonpoint Source Pollution. M. R. Overcash and J. M. Davidson (Eds.). Ann Arbor Science, Ann Arbor, MI, pp. 69–98.

White, R. E. 1988. Leaching. In: Advances in Nitrogen Cycling in Agricultural Ecosystems. J. R. Wilson (Ed.). Commonwealth Agricultural Bureau International, Wallingford, England, pp. 193–211.

Wight, J. R. (Ed.). 1983. SPUR—simulator of production and utilization of rangelands. A rangeland model for management and research. Misc. Publ. No. 1431. U.S. Department of Agriculture, Washington, D.C. p. 120.

Williams, J. R. 1975a. Sediment routing for agricultural watersheds. *Water Res. Bull.* 11(5):965–974.

Williams, J. R. 1975b. Sediment yield prediction with universal equation using runoff energy factor. In: Present and Prospective Technology for Predicting Sediment Yield and Sources. ARS-S-40. Agricultural Research Service, U.S. Department of Agriculture, Washington, D.C., pp. 244–252.

Williams, J. R. 1978a. A sediment yield routing model. In: Proc. of the Speciality Conference on Verification of Mathematical and Physical Models in Hydraulic Engineering. American Society of Civil Engineers, College Park, MD.

Williams, J. R. 1978b. A sediment graph model based on instantaneous unit sediment graph. *Water Resour. Res.* 14(4):659–664.

Williams, J. R. 1980. SPNM, a model for predicting sediment, phosphorus, nitrogen yields from agricultural basins. *Water Resour. Bull.* 16(5):843–848.

Williams, J. R. 1983. EPIC, The Erosion-Productivity Impact Calculator, Volume 1: Model Documentation. Agricultural Research Service, U.S. Department of Agriculture, Washington, D.C..

Williams, J. R. and H. D. Berndt. 1976. Sediment yield prediction based on watershed hydrology. Paper No. 76-2535. American Society of Agricultural Engineers, St. Joseph, MI.

Williams, J. R. and R. W. Hann. 1978. Optimal operation for large agricultural watersheds with water quality constraints. Technical Report No. 96. Texas Water Resources Institute, Texas A & M University,

Williams, J. R. and W. V. LaSeur. 1976. Water yield model using SCS curve numbers. *J. Hydraul. Div. Am. Soc. Civ. Eng.* 102:1241–1253.

Williams, J. R. and A. D. Nicks. 1982. CREAMS hydrology model—Option 1. In: Applied Modeling in Catchment Hydrology, Proc. of the International Symposium on Rainfall-Runoff Modeling. V. P. Singh (Ed.). Water Resource Publications, Littleton, CO, pp. 69–86.

Williams, J. R., P. T. Dyke, and C. A. Jones. 1983. EPIC: a model for assessing the effects of erosion on soil productivity. In: Analysis of Ecological Systems: State-of-the-Art in Ecological Modelling. W. K. Lauenroth et al. (Ed.) Elsevier, Amsterdam, pp. 553–572.

Williams, J. R., A. D. Nick, and J. G. Arnold. 1985. Simulator for water resources in rural basins. *J. Hydraul. Eng.* 111(6):970–986.

Wilson, B. N., B. J. Barfield, and I. D. Moore. 1982. A hydrology and sedimentology watershed model. 1. Modeling techniques. Department of Agricultural Engineering. Special Publication. University of Kentucky, Lexington, KY.

Wilson, B. N., B. J. Barfield, and R. C. Warner. 1986. Simple models to evaluate nonpoint pollution sources and control. In: Agricultural Nonpoint Source Pollution. *Dev. Environ. Model.* 10:231–263.

Wilson, J. R. (Ed.). 1988. Advances in Nitrogen Cycling in Agricultural Ecosystems. Commonwealth Agricultural Bureau International, Wallingford, England.

Winteringham, F. P. W. 1977. Joint FAO/IAEA/Federal Republic of Germany International Co-ordination program of research on agricultural nitrogen residues with particular reference to their conservation as fertilizers and behaviour as potential pollutants. *Prog. Water Technol.* 8:179–187.

Wischmeier, W. H. and D. D. Smith. 1965. Predicting Rainfall-Erosion Losses from Cropland East of Rocky Mountains — Guide for Selection of Practices for Soil and Water Conservation. Agricultural Handbook No. 282. U.S. Department of Agriculture, Washington, D.C.

Wischmeier W. H. and D. D. Smith. 1978. Predicting rainfall erosion losses. Agricultural Handbook No. 537 U.S. Department of Agriculture, Washington, D.C.

Woolhizer, D. A. and J. A. Liggett. 1967. Unsteady one-dimensional flow over a plane — the rising hydrograph. *Water Resour. Res.* 3:753–771.

Yalin, M. S. 1963. An expression for bed-load transportation. *J. Hydrol. Div. Am. Soc. Civ. Eng.* 89(HY3):221–250.

Yang, C. T. 1972. Unit stream power and sediment transport. *J. Hydraul. Div. Am. Soc. Civ. Eng.* 98(HY10):1805–1826.

Yang, C. T. 1973. Incipient motion and sediment transport. *J. Hydraul. Div. Am. Soc. Civ. Eng.* 99(Hy10):1679–1704.

Yang, C. T. 1976. Minimum unit stream power and fluvial hydraulics. *J. Hydraul. Div. Am. Soc. Civ. Eng.* 102(HY7):919–934.

Young, R. A. and C. A. Onstad. 1976. Predicting particle size composition of eroded soil. *Trans. ASAE* 19(6):1071–1075.

Young, R. A., C. A. Onstad, D. D. Bosh, and W. P. Anderson. 1985. Agricultural Nonpoint Source Pollution model (AGNPS) I and II. Model documentation. Minnesota Pollution Control Agency, St. Paul, MI, and Agricultural Research Service, U.S. Department of Agriculture, Washington, D.C.

Young, R. A., C. A. Onstad, D. D. Bosch, and W. P. Anderson. 1987. AGNPS, Agricultural nonpoint source pollution. A watershed analysis tool. Conservation Research Report 35. U.S. Department of Agriculture, Washington, D.C.

Zison, S. W. 1980. Sediment-pollutant relationships in runoff from selected agricultural suburban and urban watersheds: a statistical correlation study. EPA-600/3-80-022. U.S. Environmental Protection Service, Washington, D.C.

CHAPTER 3

Modeling Contaminant Transport in Subsurface: Theory and Computer Programs

D. A. Barry

3.1 INTRODUCTION

The purpose of this chapter is to discuss the theory of chemical transport in soil, as well as available computer codes for implementing that theory. Chemical transport in the subsurface flow is generally a complex process. Mathematical models that describe contaminant transport can involve substantial contributions from the basic disciplines of mathematics, biology, physics, and chemistry. The interdisciplinary nature of contaminant transport modeling requires that it be a synthesis of disciplinary components. Accurate predictions of chemical transport phenomena are required in many different domains, spanning hydrology, environmental engineering, soil science, and agriculture. The circumstances in which a model may be needed vary widely as well. At the local scale, one might model a leachate plume moving from a landfill into a groundwater system, while at a much larger scale one might consider regional salinization leading to degradation of both water supplies and soil. The unifying theme is that a basic theoretical structure can be applied in each case. That is, besides modeling contaminant transport per se, the same basic theoretical approach applies to a wide range of processes that include both geochemistry and transport, e.g., weathering, diagenesis, hydrothermal ore deposition, geothermal energy production, and oil recovery (Lichtner, 1985). Of course, all these applications require different manifestations of the basic theory.

In this chapter we review the theory underlying the transport models that are used. The contaminant transport subdiscipline of hydrology and the geochemistry

ISBN 0-87371-747-3

subdiscipline of hydrogeology seem to be separated by a gulf of both time and spatial scales in the processes that they consider. This gulf is most highlighted when the crustal flow problem is considered. Torgensen (1991), reporting on the scientific approaches taken to this problem, suggests that the hydrologic view is that large-scale fluid flow in the crust is essentially a continuum process in sedimentary basins. These processes are driven by topography and additional gradients in fluid density caused by temperature and solute concentration. Chemical reactions in such systems are rare, and can be ignored. The geological viewpoint, which Torgensen (1991) calls the structural/metamorphic view, is that fluid flow in the crust is highly episodic and controlled by rock structural features such as fractures. Chemical reactions are thought to be poorly understood but very important. The difference between these approaches is mainly one of temporal and spatial scales and emphasis. The hydrologic viewpoint is appropriate for assessing anthropogenic concerns at the regional scale or below. The structural/metamorphic view applies to much larger scales, both in time and space. An appropriate spatial scale might be, for example, a continental or subcontinental scale. Australia's Great Artesian Basin, which underlies a large portion of the eastern half of the continent, represents the scale appropriate to the structural/metamorphic view. On the other hand, the hydrologists' view is the one adopted to determine a yearly water balance on a single catchment. We adopt the perspective of the hydrologist, but consider that both physical and chemical processes need be included in developing models of solute transport in soil. To adopt the rather humorous phrasing of Torgensen (1991), the soil is not a "Teflon box" in which chemical reactions can be ignored. For that reason some emphasis is placed on elucidating the underlying chemical and physical models that are the basis of flow and transport computer codes. Consequently, to have some appreciation of the chemical reaction and speciation modeling that is used, relevant chemistry concepts and theory are discussed as well. The modeling approaches are discussed in the text, while basic chemistry concepts are presented in the appendix (Section 3.10). Mathematical formulations of speciation and transport models are presented with the aim of indicating the diverse approaches one can take in formulating and solving a governing model. Computer modeling efforts have been divided into three categories:

1. Geochemical models
2. Multicomponent transport models
3. Semianalytical transport models

Models in category 1 consider the chemical reactions but ignore flow and transport. Those in category 2 incorporate category 1 models into a flow and transport model to produce a comprehensive ("hydrologic") description of solute movement and reactions in soil. Category 3 models are mostly single-species transport models that include approximations for reactions, ion exchange, and/or solute sorption. We begin by a brief discussion of the problem of determining equilibrium speciation in aqueous solutions. Many of the existing models are based on decoupling of water flow and solute transport, particularly for single-species transport. The codes devoted to flow only are not included here. Mangold and Tsang (1991) provide a valuable overview of these and other available programs.

3.2 REACTIONS IN AQUEOUS SOLUTIONS: DETERMINATION OF SPECIATION AT EQUILIBRIUM

Chemicals in aqueous solution may or may not react with each other. In natural soil environments there may be many chemical species in solution at any one time. Speciation models have been developed to determine the composition of the solution at equilibrium in terms of the constituent species. The total set of species can be defined by a set of equations involving components. Components are chemical entities that are indivisible for the purpose of the speciation calculations. Thus, components can be ions or complexes, for example. Coves and Sposito (1986, §1.1) point out — to use the language of functional analysis — that the components form a basis for the linear (species) space (Brown and Page 1970, §3.1). Before any speciation calculations can begin, a component equation for each chemical species must be written. Each species is written as the result of a chemical reaction involving only the components. The relationship between the species and components is quantified using the law of mass action (see the appendix). If there are N species, N equations of the following form can be derived (Coves and Sposito, 1986, §1):

$$\left\{c_i\right\} = K_{e,i} \prod_{j=1}^{M} \left\{X_j\right\}^{a_{i,j}}, \quad i = 1,\ldots,N \tag{1}$$

The activities in Equation (1) may be replaced by molar concentrations for dilute solutions. If there are no sources or sinks of species in the system then the total mass of each component must be fixed, i.e.,

$$X_{T,j} = \sum_{i=1}^{N} a_{i,j} c_i, \quad j = 1,\ldots,M \tag{2}$$

The aim is to compute the equilibrium concentrations, c_i, of the N species in solution. Because each species is written in terms of components, the problem is equivalent to computing the concentration of each component. The total concentration of each component, $X_{T,j}$, $j = 1, \ldots, M$, is known, so Equation (2) gives M equations to find the M unknown component concentrations. Of course, the species concentrations in the summation of Equation (2) are unknown, but they are given in terms of the individual components by Equation (1), because the equilibrium coefficients are assumed known. Thus, Equations (1) and (2) define a nonlinear system of equations from which the component concentrations can be found numerically. Most computer codes use Newton's method (e.g., Bajpai et al., 1977, Ch. 4) for this purpose. Finally, the concentration of each species in the system — the object of the exercise — is calculated from the component concentrations.

Mineral precipitation or dissolution can be handled with a minor variation to this process. The equilibrium constant is replaced by the solubility product, K_{sp}. From Equation (A7) we obtain:

$$K_{sp} \geq \{R_1\}^{r_1} \times \cdots \times \{R_N\}^{r_N} \tag{3}$$

where the denominator in Equation (A7) has been replaced by unity, which is the activity of a precipitate. If Equation (3) holds, then the precipitate will be formed and the reaction needs to be accounted for. Conversely, it can be ignored when Equation (3) does not hold. Morel and Morgan (1972), Reed (1982), Walsh et al. (1984), Wigley (1977), and Kirkner and Reeves (1990) all present computational algorithms for including such reactions.

The procedure outlined here to calculate the equilibrium speciation of a closed system is but one method to achieve this goal. Other methods can be derived and, incidentally, traced to Brinkley (1946, 1947), but they contain the same essential elements as those mentioned above (Sposito, 1984a, §2). The key factors are the development of the chemical equations for the system and knowledge of the equilibrium constants for the various reactions. The former must characterize the major reactions in the system while the algebraic accuracy of solution depends, in part, on the latter.

3.3 SOLUTE ADSORPTION AND EXCHANGE

Solute adsorption in soil occurs when solute is removed from the transporting water and is collected at sites within the solid porous medium at the interface of the solid and liquid phases (Tan, 1982, §4.1). For a particular soil sample that is mixed with a given mass of fluid of known initial concentration, it is clear that the concentration of solute in the fluid may increase or decrease depending on the chemical status of the soil. For example, Sposito (1989, §8.1) considers an experiment in which 1 kg of dry soil is combined with M_{Tw} kg of water containing a solute at a specified molality. The excess, q, of the chemical species is defined as

$$q = \Delta m M_{\mathrm{Tw}} \tag{4}$$

where Δm is the change in molality, relative to the original composition of the solution, of the soil due to adsorption. Thus, if the soil initially has no net charge and solute is adsorbed onto the soil then $q < 0$ and vice versa. For experiments carried out at constant temperature, it is permissible to measure concentration of the solute in terms of molarity (Sposito, 1981, §2.2), c, instead of molality. Experiments are usually carried out under isothermal conditions. In such experiments, one measures the quantity of solute adsorbed per unit mass of dry soil, s, for various predetermined concentrations of the pore fluid. The experimental results consist of data pairs (s_i, c_i) that, when plotted, define an adsorption isotherm. If, for example, the c_i data are measured in units of moles per cubic meter of soil solution (i.e., pore fluid), while the s_i data are measured in units of moles per kilogram of dry soil, then $w_i = s_i\rho_b/\theta$ gives the adsorbed concentration in moles per cubic meter of soil to solution. In this review we focus only on rapid sorption, i.e., the isotherms do not have an explicit time dependence. A readable discussion of the latter topic is provided by Barrow (1989).

3.3.1 Adsorption Isotherms

The two most common isotherms used in practice to describe data from adsorption experiments are the Langmuir and Freundlich isotherms, given by Equations (5) and (6) in Table 3.1, respectively. Numerous derivations of the Langmuir isotherm are available (Young and Crowell, 1962, Ch. 4), including one based on statistical mechanics (Sposito, 1979). It is possible also to derive the Freundlich isotherm (Sposito, 1980) mechanistically, although Jaycock and Parfitt (1981, §5.3) state that it is "essentially empirical, and perhaps only useful for fitting data." Indeed, referring

Table 3.1 Adsorption Equations

Isotherm	Name	Ref.	Equation No.
$w = \dfrac{k_1 c}{1 + k_2 c}$	Langmuir	Langmuir (1918)	(5)
$w = k_3 c^{k_1}$	Freundlich	Freundlich (1909, 1926, p. 172)	(6)
$w = \dfrac{k_1 c}{1 + k_2 c} + \dfrac{k_4 c}{1 + k_5 c}$	Double Langmuir	Shapiro and Fried (1959)	(7)
$w = \dfrac{k_1 c}{1 + k_2 c + k_6 \sqrt{c}}$	Gunary	Gunary (1970)	(8)
$w = k_3 c^{k_4} - k_7$	Fitter-Sutton	Fitter and Sutton (1975)	(9)
$w = \dfrac{k_8 c^{1-k_1}}{1 + k_9 c^{1-k_1}}$	Extended Langmuir	Sibbesen (1981)	(10)
$w = k_{10} c^{c^{k_1/k_8}}$	Extended Freundlich	Sibbesen (1981)	(11)
$w = k_7 \left\{1 - \left[1 + (k_2 c)^{k_1}\right]^{k_4}\right\}$	Langmuir-Freundlich interpolation	This work	(12)

to phosphate adsorption, Sibbesen (1981) suggests that all isotherms be treated as being empirical due to the observation that "soils are complex with surfaces of mostly unknown heterogeneity."

As already stated, adsorption in soils refers to the chemical species accumulating at the surface of the solid matrix. On the other hand, absorption occurs when the species accumulates within the matrix. These two processes are not readily distinguishable in soils, so one refers instead to sorption, which includes both cases. In Table 3.1 various adsorption equations have been listed (cf. van Genuchten and Cleary, 1982, Table 10.1). Note that these isotherms do not account for time dependence of adsorption, i.e., adsorption kinetics; they account only for systems in equilibrium. Table 3.1 does not exhaust the possible adsorption isotherms that may be employed. The efficacy of various isotherms in fitting data has been compared for a variety of isotherms by Kinniburgh et al. (1983) and Kinniburgh (1986). Kinniburgh (1986) and Barry (1990a) point out that some common graphical techniques of fitting isotherms to data will lead to erroneous parameter estimates. Graphic methods involve transforming the experimental data such that straight line fits can be employed. This method is prone to error, as accurate parameter estimates will result only when the transformed data are weighted appropriately.

Sposito (1982) demonstrates that the double Langmuir isotherm can be used to describe any adsorption isotherm for which the distribution coefficient is smooth, concave, and tends asymptotically to a finite limit. As mentioned above, most isotherms are empirical, but some are based on foundational theories. Such theories provide a sufficient, but not a necessary, condition for the adsorption mechanism. One cannot reach any conclusions concerning the adsorption mechanism just from the ability of a particular isotherm to describe the experimental data; other, ancillary information is needed. Many of the equations in Table 3.1 have been compared by Ratkowsky (1986) using data collected by Holford (1982). The Freundlich isotherm and its extension, Equation (11), were found to be the best statistical descriptions of the data. Therefore, if an adsorption isotherm is used merely as a convenient description of data, one can choose the most accurate mathematical description available from Table 3.1 or elsewhere. It must be recognized that the adsorption predicted by the Freundlich-type isotherms is not bounded as concentration increases, and so extrapolation based on a fitted isotherm may describe poorly the case being modeled.

Sposito (1989, Figure 8.1) identifies four types of adsorption isotherm, namely the S-, L-, H-, and C-curve, and discusses possible mechanisms underlying each one. Qualitative depictions of these types have been included in Figure 3.1. The designations of the curves come from their behavior at relatively low concentrations (Sposito, 1981, §5.2). If one considers only the two most common isotherms, the Langmuir and Freundlich isotherms, then it can be shown that either or both isotherms have the form of the latter three curves. However, neither isotherm can reproduce the S-curve. Equation (12) in Table 3.1 is a proposed new empirical isotherm. It is suggested because it reduces to an interpolation between the Langmuir and Freundlich isotherms. Additionally, it can reproduce S-curves without difficulty. First, consider Equation (12) with $k_1 = -k_4 = 1$:

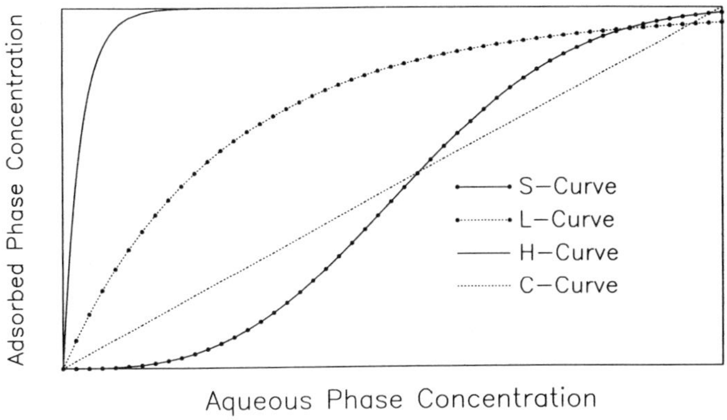

FIGURE 3.1. General classes of adsorption or exchange isotherms. S-curve (●—●): relatively low affinity of the exchangeable ion for the exchanger; L-curve (●···●): relatively high affinity of the exchangeable ion for the exchanger; H-curve (—): extremely high affinity of the exchangeable ion for the exchanger; C-curve (···): linear exchange isotherm. Equation (12) can produce each class of isotherm.

$$w = k_7\left(1 - \frac{1}{1 + k_2 c}\right) = \frac{k_7 k_2 c}{1 + k_2 c} \tag{13}$$

If we take $k_7 = k_1/k_2$ then Equation (13) reduces to Equation (5). If, instead, we set $k_4 = 1$ then Equation (12) reduces to

$$w = k_7\left(1 - 1 + k_2^{k_1} c^{k_1}\right) = k_7 k_2^{k_1} c^{k_1} \tag{14}$$

Equation (14) is equivalent to Equation (6) if $k_7 k_2^{k_1} = k_3$. That Equation (12) produces S-curves can be confirmed by comparison with the characteristic S-shaped water content-pressure head function proposed by van Genuchten (1979, [3]). Equation (12) can be seen to be a variation of van Genuchten's function. To demonstrate the utility of Equation (12), we fit it to some experimental data extracted from Figure 4.1 of Sposito (1984c), as shown in Figure 3.2. The data show copper adsorption by the Altamont clay loam. This figure shows that w, as predicted by Equation (12), is constrained to lie between 0 and k_7. However, the experimental data shown in Figure 3.2 suggest a nonzero value of w near $c = 0$. The isotherm in Equation (12) could easily be scaled to pass through this nonzero datum, and so improve the fit to these data.

3.3.2 Ion Exchange

Thus far, no mention has been made of the ions displaced during the sorption process. If different ions are being displaced then ion exchange must be considered. For example, a generic cation exchange reaction is

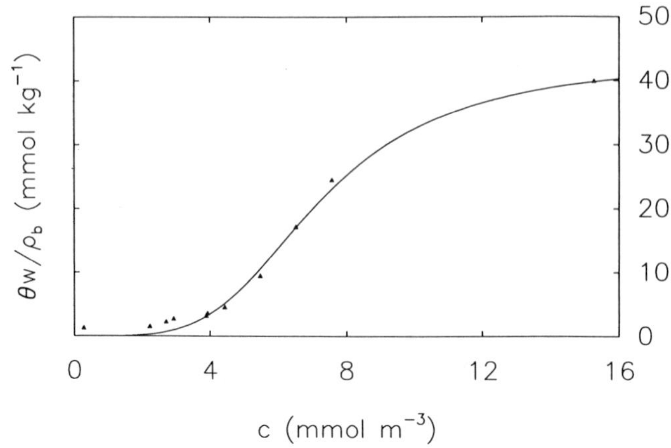

FIGURE 3.2. Plot of the adsorption isotherm [Equation (12)] and experimental data (Sposito, 1984c, Figure 4.1) showing copper adsorption onto the Altamont clay loam. Symbols: ▲, experimental data; —, Equation (12) with the parameter values k_1 = 4.411 and k_2 = 0.161 mmol^{-1} m^3, k_4 = –0.627 and k_7 = 43.474 mmol m^{-3}.

$$gFY_f(s) + fGX_g(1) \rightleftarrows gFX_f(1) + fGY_g(s) \tag{15}$$

where s is the solid phase and l is the liquid phase, F and G are cations with valence f and g, respectively, and X and Y are anions, or anion complexes, with valence –1. The equilibrium constant for this reaction is defined by analogy with Equation (A7). Then, the resulting expression can be written in terms of equivalent fractions of F and G in the solid and liquid phases. Let E_F be the equivalent fraction of F in the solid phase, and $\tilde{E}_F$ be the equivalent fraction of F in the liquid phase, with similar quantities defined for cation G. The equilibrium constant for Equation (15) can then be written in terms of these equivalent fractions (Barrer and Klinowski, 1974; Sposito, 1981, §5.2). To simplify matters, let F have valence 2 and G have valence 1, and consider a reaction where the equilibrium constant has unit value, i.e., the exchanger has no preference for either cation. An equation relating the equivalent fractions in the adsorbed and liquid phases for the univalent ion G is given by (Sposito, 1981, §5.2):

$$\frac{E_G^2}{1 - E_G^2} \propto \text{TN} \frac{\tilde{E}_G^2}{1 - \tilde{E}_G} \tag{16}$$

where TN is the total normality of the liquid phase. We note that the proportionality factor in Equation (16) varies with TN. We note that the simple algebraic form of Equation (16) is due to the valences chosen for each species. Equation (16) serves, however, to illustrate two points: first, that the amount of a single species in the solid and liquid phases can be related by a single expression and, therefore, can be used to compute isotherms because equivalent fractions can be converted easily to concentration units. In many cases, the adsorption isotherms presented above are general enough to reproduce these curves empirically. Second, these isotherms can change as the chemical conditions of the system change. In Equation (16) the relationship between E_G and $\tilde{E}_G$ varies from nearly linear to strongly nonlinear depending on the value of TN. So far, the solid phase on which the exchange is occurring has been assumed to be homogeneous. Soil can contain many different classes of exchange sites, and theory accounting for exchange reactions in such situations has been developed (Barrer and Klinowski, 1979).

3.4 CONTINUUM MODEL OF MULTICOMPONENT, AQUEOUS-PHASE SOLUTE TRANSPORT

Until this point only basic aqueous-phase chemical reactions have been discussed. Each chemical component in the porous medium is also advected and dispersed by the flowing fluid. The starting point for predicting solute concentration in the subsurface is the construction of a model describing both the physical processes induced by the fluid flow field as well as the chemical reactions that are occurring. Only with such a model can the effects of combined physical and chemical processes be quantified. Such a model is presented here, although a more detailed presentation is available elsewhere (Kirkner and Reeves, 1988). We point out that the mathematical basis for the model presented is that the porous medium and the fluid constituents can be replaced by a continuum or, rather, a set of continua that interact over the spatial domain of interest. Foundational theories leading to continuum models (e.g., Hassanizadeh and Gray, 1979a,b; Marle, 1982) are based on the concept of a representative elementary volume, denoting a local averaging scale for the porous medium. It is assumed here that a continuum modeling approach is valid.

The basic equations governing the flow of N_c chemical components in a flowing system are (Bear, 1972; Bryant et al., 1986)

$$\frac{\partial\left(\theta c_i^T\right)}{\partial t} + \mathbf{L}\left(u_i\right) = 0, \quad i = 1, \ldots, N_c \tag{17}$$

where c_i^T is the *total* concentration of the *i*th component in any form whatsoever in the system and u_i is the concentration of the same component in the *flowing* (aqueous) solution. The right side of Equation (17) is nonzero if production terms need be included, or if irreversible dissolution or precipitation occurs. The operator $\mathbf{L}$ is defined by (Kirkner and Reeves, 1988)

$$\mathbf{L}\left(u_i\right) = div\left(u_i\mathbf{q} - \theta\mathbf{D}\ \mathrm{grad}u_i\right) \tag{18}$$

Now, the total concentration of the *i*th component in the system is, by definition,

$$\underset{\text{Total concentration}}{c_i^{\mathrm{T}}} = \underset{\text{Component concentration}}{c_i} + \underset{\text{Sorbed concentration}}{s_i} + \underset{\text{Complexed concentration}}{\sum_{j=1}^{N_x} A_{ij} x_j} \tag{19}$$

In Equation (19) it has been assumed that there is no precipitation or dissolution taking place. For conditions where precipitation can occur, an additional precipitation term must be included on the right side of Equation (19). In terms of equilibrium chemical processes, the inclusion of such a term does not present any difficulties. For solute transport within a permeable medium, the inclusion of such a term is a complicating factor, because porosity will vary as precipitation or dissolution occurs. On the other hand, if the quantity of precipitation is such that porosity changes are not significant then the usual approach is to model precipitation as a kind of aqueous complexation. Indeed, the only change is that instead of an equilibrium constant applying to the reaction, one considers the solubility product for the particular precipitate. If the solubility product is exceeded then precipitation takes place, else dissolution occurs.

There have been models developed that explicitly account for porosity changes accompanying precipitation (Helgeson et al., 1984; Lasaga, 1984; Lichtner, 1987). The impact of precipitation is often more important under geologic time scales. Typical contaminant transport operates at much shorter time scales with relatively dilute chemical concentrations, so precipitation effects are not considered further here. We recognize, however, that such phenomena need be included in general models.

The latter two terms on the right side of Equation (19) are now discussed. In the case of soluble complexation, the appropriate chemical reactions are of the generic form (Kirkner and Reeves, 1988):

$$\sum_{j=1}^{N_x} A_{ij} \hat{c}_j \rightleftarrows \hat{x}_i;\ K_i, \quad i = 1, \ldots, N_x \tag{20}$$

where K_i is the equilibrium constant for the reaction and the circumflex indicates a chemical formula, not a concentration. Upon using the law of mass action (see the appendix), one obtains from Equation (20):

$$x_i = K_i \prod_{j=1}^{N_c} c_j^{A_{ij}}, \quad i = 1, \ldots, N_x \tag{21}$$

If the concentrations in Equation (20) are not dilute, then each concentration variable must be replaced by the activity of each species. The total soluble concentration of the ith component, u_i, is defined by

$$u_i = c_i + \sum_{j=1}^{N_x} A_{ij} K_j \prod_{k=1}^{N_c} c_k^{A_{jk}}, \quad i = 1, \ldots, N_c \tag{22}$$

Equation (22) shows that the soluble concentration can be specified as a nonlinear function of the component concentration variables alone. For later use, we observe that Equation (22) defines N_c unknowns, u_i, in terms of N_c (presumed) known variables, c_i. In principle then, Equation (22) can be inverted to give each c_i as a function of the set of N_c u_i variables, i.e. (Kirkner and Reeves, 1988):

$$c_i = f_i^{cu}\left(u_1,\ldots,u_{N_c}\right), \quad i-1,\ldots,N_c \tag{23}$$

Previously, adsorption isotherms were discussed in some detail. The main point to note here is that it is always possible to write the adsorbed concentration, s_i, in terms of the fluid phase solute concentration, i.e.,

$$s_i = f_i^{sc}\left(c_1,\ldots,c_{N_c},x_1,\ldots,x_{N_x}\right) \tag{24}$$

Equation (24) becomes a function of the component concentrations, c_i, if the complexed concentrations are eliminated using Equation (21). The form of the function in Equation (21) is more general than the isotherms presented in Table 3.1. In Table 3.1, adsorption was considered for a single species only, whereas in Equation (24) the adsorption of any component depends on the concentration of all species in solution. The isotherms in Table 3.1 can be extended to account for different species in solution. For example, Kirkner and Reeves (1988) provide a simple derivation of the multicomponent Langmuir isotherm (Butt, 1980):

$$s_i = \frac{c_i K_{s_i} \mathrm{x_T}}{1+\sum_{j=1}^{N_c} c_j K_{s_j}} \tag{25}$$

The governing Equation (17) can be cast with either c_i or u_i as the dependent variable using Equations (18) through (24). This means any numerical solution of the multicomponent chemical transport model must solve N_c-coupled, nonlinear partial differential equations. Clearly, even moderate-sized cases of this class of problem represent a challenge for even the most powerful supercomputer (Barry, 1990b).

There are a variety of ways in which the governing model of reactive chemical transport can be tackled numerically. Kirkner and Reeves (1988) detail some analytical results, while Reeves and Kirkner (1988) provide case studies solved using different numerical schemes. There is, apparently, no single numerical scheme that is most efficient for all cases, i.e., the efficiency of the scheme is dependent on chemical characteristics. Perhaps more interesting are the results of the various cases considered by Reeves and Kirkner (1988). Reactions that involve relatively high rates of complexation or adsorption represent the most difficult numerical challenge. Such cases also lead to the most dramatic rates of formation or disappearance of species within the porous medium. For instance, chemical reactions may lead to the presence of relatively high concentrations of a particular species within a given soil volume, whereas at the boundaries where fluxes are measured, the species concentration may be low. Thus, assumptions concerning the concentration of species within a medium, based on surface measurements, may be erroneous if the chemical reactions taking place favor the production of that species.

The flow of the transporting fluid has not been considered yet. Fluid flow in the unsaturated zone is governed by Richards' equation (Richards, 1931), while in the saturated zone it is a parabolic equation mathematically similar to Equation (17). Because the Darcy flux is dependent on the density of the fluid, the chemical transport equations, Equation (17), and the flow equation are coupled through the density equation of state. Similarly, if the transport of heat is important then the appropriate governing model is again similar in form to Equation (17). Temperature changes induce density changes in the fluid, so again there is coupling between the equations. The purpose to which the model is to be put is the determining factor to be used in model formulation and selection.

The theory used for quantifying chemical reactions and transport in porous media has been outlined and the main features of the theory demonstrated. The development of computer codes for chemical reactions and transport varies markedly in scope and complexity. Geochemical codes, for instance, are usually limited in usefulness by their accompanying databases of thermodynamic information. Transport codes for chemical species and fluid movement are likewise subject to a variety of numerical approaches. Many specialized geochemical and transport and flow codes are available. The nature of multicomponent chemical transport models is such that it is convenient to combine available codes for geochemistry and for flow and transport into a complete model. Below we look sequentially, therefore, at the available geochemical codes and then at existing multicomponent transport codes. The latter quantify the combined effects of the physics of flow and the chemical reactions of the transported species. As stated at the outset of this chapter, the variety of chemical transport scenarios is enormous. We describe also less complex computer codes that are used to simulate more simplified cases of transport, usually with correspondingly simplified chemical processes. To begin with, geochemical computer codes are discussed.

3.5 GEOCHEMICAL MODELS

3.5.1 Major Computer-Based Geochemical Codes

Many geochemical computational codes have been developed. Over time, code developments have continued, and improved versions of the various codes have been released. Jenne (1981) provides an overview of such developments. A list and brief description of major computer-based geochemical codes is presented in Table 3.2. This list is not meant to be exhaustive, rather it indicates the wide spectrum of available programs. Some of these codes are described in more detail below.

Table 3.2 Summary of Computer-Based Geochemical Models[a]

Name	Description	Ref.
CHEMIST	CHEMIST uses the Gibbs free energy minimization method to compute the equilibrium concentration distribution in a closed system. Alternatively, the steady state distribution in an open system can be determined	Deland (1967)
CHILLER	Computes multicomponent heterogeneous equilibria in all phases considering variations in temperature, pressure, and enthalpy	Reed (1991)
EQQYAC	EQQYAC uses the ion-association model. Activity coefficients depend on total concentration (assumed less than 1 molal). Thermodynamic data include temperature-dependent equilibrium constants	Barragan and Nieva (1989)
EQUILIB	Equilibrium concentrations between gaseous, aqueous, and solid phases are calculated. Aqueous speciation, redox, and pH can be determined in either open or closed systems	Morrey (1981)
EQ3/6	Computes distributions of ions, neutral species, ion pairs, and complexes in natural waters using a database. Irreversible reactions in solid rock systems and the effects of heating and cooling of solutions are included	Wolery (1979)
GEOCHEM	A modification of REDEQL-2, this code is able to compute chemical equilibria in natural soil solutions, including precipitation/dissolution, redox reactions, and adsorption. It includes a database	Sposito and Mattigod (1980)
MINTEQ	MINTEQ allows computation of ion speciation, activity coefficients, adsorption, and mass transfer. It is accompanied by a well-documented database. It is a partial descendent of WATEQ	Felmy et al. (1984)
PENALTY	The precipitation/dissolution reaction portion of the equilibrium speciation problem is handled with a penalty function method	Kirkner and Reeves (1990)

Table 3.2 Summary of Computer-Based Geochemical Models[a] (continued)

PHREEQE	This is an evolution of WATEQ and MIX2. Equilibrium calculations are based on aqueous phase concentrations and charge neutrality. No mass constraints are imposed on hydrogen and oxygen	Parkhurst et al. (1980)
SOLMNEQ	Calculates speciation for 162 inorganic aqueous phase species found in natural waters in the temperature range 0–350°C	Kharaka and Barnes (1973)
SOLVEQ	Computes multicomponent homogeneous equilibria of aqueous chemical species. For a known chemical composition and temperature, aqueous species activities, saturation indices, and gas fugacities are calculated	Reed (1982)
STEADYQL	Chemical speciation and composition in soil-water systems are calculated for reactions at three time scales: very fast reactions (equilibrium model), moderate (kinetic equations), and very slow (no reaction at the time scale of interest)	Furrer et al. (1989)
WATEQX	A modification of WATEQF that generalizes the program by extending the ionic strength range and updates the database using a commercial database program	van Gaans (1989)
WATEQ3	This is part of a series of improvements to the original WATEQ model, and includes an extensive thermodynamic database. Equilibrium distributions in natural waters are calculated with effects of temperature, pH, and redox potential included	Ball et al. (1981a,b)

[a] Cf. Kincaid et al. (1984a,b).

3.5.2 Descriptions of Some Geochemical Codes

3.5.2.1 EQ3/6

This package consists of two main parts: EQ3NR and EQ6, together with a shared thermodynamic database. EQ3NR is a speciation-solubility code that models the state of an aqueous solution. Reaction paths are calculated by EQ6, wherein aqueous systems are modeled as they move toward equilibrium, for example, reaction paths for reactions modeled by kinetic rate laws. Two scenarios are modeled by EQ6: (1)

a closed system and (2) a parcel of fluid traversing a reactive medium. Case (2) represents a crude transport model. EQ3/6 has been used to model tuff-groundwater interactions (Kerrisk, 1984) and crust and sea-water interactions (Bowers et al., 1985).

3.5.2.2 EQQYAC

This code consists of a main program and four subroutines (Barragan and Nieva, 1989). One subroutine computes temperature-dependent activity coefficients using Debye-Hückel theory (Henley et al., 1984). Another contains expressions for mass action equilibrium constants (Giggenbach, 1980). The subroutine ENTHALPIA computes specific enthalpy of steam and liquid phases (Fournier and Potter, 1982) while the final subroutine determines the steam fraction using the Fisher-Tropsch reaction (Nieva et al., 1987). EQQYAC has been used to calculated equilibrium concentrations of inorganic ions in a geothermal reservoir. CHILLER, which uses the data base SOLTHERM (Reed and Spycher, 1989), is another geochemical code that was designed for geothermal reservoir modeling. In most nongeothermal applications, such extensive temperature-dependent data sets will not be necessary.

3.5.2.3 GEOCHEM

This code is specifically aimed at calculating equilibria in soil solutions. Both redox reactions and solid phases can be included or excluded, and exchange coefficients can be estimated. The user can choose to have variable pH and ionic strength. Considerable effort has been devoted to compiling the thermodynamic database that draws on the work of Martell and Smith (1976–1977), Baes and Mesmer (1976), Robie et al. (1978), and Sadiq and Lindsey (1979), although the authors emphasize that there remains the possibility of uncertainty in the data. Parker et al. (1988) used a modified version of GEOCHEM to determine aluminium speciation in a soil solution with the results used to indicate aluminum toxicity to plants. They found, for example, that the activity of Al^{3+} has a detrimental effect on root growth in wheat. GEOCHEM was evaluated by Kincaid et al. (1984b), who concluded that the code and its accompanying documentation is very well suited to users who are disinclined to modify the code. On the other hand, documentation was not suited to those whose applications required that modifications be made. GEOCHEM has recently been updated and renamed SOILCHEM (Sposito and Coves, 1988).

3.5.2.4 MINTEQ

MINTEQ computes chemical equilibria in aqueous solutions, gases, and solids. The thermodynamic database contains information on numerous elements and complexes. It can model adsorption by the linear, Freundlich, or Langmuir isotherms. Alternatively, surface complexation models can be used, e.g., the constant capacitance or triple layer models. Surface complexation models are based on an assumed model of ion behavior at the local level (Sposito, 1984b, Ch. 5). In MINTEQ, the Davies and Debye-Hückel equations are used to calculate activity coefficients for aqueous species. The nonlinear set of equations set up by MINTEQ are solved using the Newton-Raphson procedure. The model has been used to simulate reactions in leaching of mining wastes (Peterson and Krupka, 1981).

A reduced version of MINTEQ, called MININR, has been produced by Felmy et al. (1984) for incorporation in flow and transport models. In essence, MININR is MINTEQ without the user interface. MININR was combined with a simple model for saturated flow by Felmy et al. (1984). This flow, transport, and chemical reaction model was tested against data from a laboratory column experiment where water was percolated through retorted oil shale. The combined model was able to predict with reasonable accuracy the pH, and concentrations of Ca^{2+}, SO_4^{2-}, Na^+, and K^+.

3.5.2.5 WATEQX

WATEQX is a version of the WATEQ family (e.g., Garrels and Thompson, 1962; Truesdell and Jones, 1974; Plummer et al., 1976). This family is used to find equilibrium speciation in natural waters using mass balance on all components except the solvent. The pH is a required input parameter, as is the temperature and redox potential. The main change incorporated in WATEQX is that data for reactions not contained in the present database can be incorporated quite easily. Activity coefficients can be computed using the extended Debye-Hückel and Davies formula, or the mean salt method of Pitzer (1973), Pitzer and Mayorga (1973), and Millero and Schreiber (1982), thereby extending the model to more electrolytes than previously possible.

3.6 MULTICOMPONENT TRANSPORT AND REACTION MODELS

3.6.1 A List of Major Codes

The number of combined chemical reaction-transport models developed remains relatively small. Many of those available have been developed at the Lawrence Berkeley Laboratory. A catalog of the major codes presently existing is given in Table 3.3. We observe that many of these codes have been developed to handle particular problems, especially those related to nuclear waste management. However, the generality of the codes, which depends in part on the geochemical data base incorporated in the model, makes them applicable, in principle, to other transport problems, e.g., transport of nonradioactive contaminants.

3.6.2 Descriptions of Some Multicomponent Chemical Transport Codes

The codes differ in the their complexity. This is displayed schematically in Figure 3.3, which considers mainly codes developed at the Lawrence Berkeley Laboratory. CHEMTRN, for example, models one-dimensional solute transport coupled with equilibrium chemical reactions. Thus, CHEMTRN can be considered to model a subset of the processes simulated by THCC because THCC includes the processes modeled by CHEMTRN in addition to heat transport. Some of the codes in Table 3.2 are described in more detail below.

3.6.2.1 CHEMTRN

This model simulates one-dimensional solute transport in a saturated porous medium. The processes of advection, dispersion, sorption, aqueous phase complex

Table 3.3 Combined Flow and Multicomponent Chemical Transport Models

Name	Description	Ref.
CHEMTRN	One-dimensional advective-dispersive transport code with equilibrium chemical reactions of speciation, precipitation/dissolution, cation exchange, and surface speciation	Miller and Benson (1983)
CHMTRNS	A nonequilibrium, temperature-dependent transport model, evolved from the CHEMTRN code	Noorishad et al. (1987)
DYNAMIX	Couples PHREEQE (Table 3.2) with the multidimensional transport code TRUMP. Now extended to include acid-base reactions, aqueous complexation, redox reactions, and precipitation/dissolution reactions	Liu and Narasimhan (1989a)
FASTCHEM	This is a suite of seven codes to model coupled flow and transport. The flow code, EFLOW, is a 2-D finite element solution of the flow equation, while the chemistry and transport codes are descendents of the MINTEQ geochemical code and the SATURN transport code	Hostetler et al. (1988)
FIESTA	Predicts the migration of subsurface pollutants that undergo convection, dispersion, soluble complexation, and sorption	Kirkner et al. (1984)
PHREEQM	Uses PHREEQE (Table 3.2) to model chemical reactions with a mixing cell concept to simulate transport	Nienhuis and Appelo (1988)
REACTRAN	Couples chemical reactions to mass transport and computes porosity changes due to precipitation and dissolution	Ortoleva et al. (1987)
SATRA-CHEM	Multidimensional flow with sorption, ion exchange, and equilibrium speciation	Lewis et al. (1986)
THCC	An extension of CHMTRNS in which heat transport is included but not coupled to mass transport	Carnahan (1987b)
TIP	Solves the coupled equations of solute, heat, and fluid transport, including equilibrium chemical reactions	Carnahan and Jacobsen (1988)

formation, and the dissociation of water are included in the model. Miller and Benson (1983) showed that the CHEMTRN predictions compared well with field data on the transport and exchange of Na^+, Ca^{2+}, and Mg^{2+} (Valocchi et al., 1983).

3.6.2.2 DYNAMIX

Liu and Narasimhan (1989a,b) consider that redox reactions are important not only in both organic and inorganic contaminant transport, but also in diagenesis of sedimentary rock, formation of oil and gas, and metal corrosion (cf. Sillen, 1967). The chemical reaction portion of DYNAMIX solves a nonlinear set of equations composed of mass balance equations, a charge balance equation, a conservation of redox state equation, and mineral solubility equations. The mineral assemblage within the system is determined by minimizing the Gibbs free energy. Oxygen is treated as a secondary, dependent species to account for changes in redox potential in the aqueous solution, and so oxygen transport is computed as part of the solution. As already stated in Table 3.3, the chemical reaction model is coupled with the transport model TRUMP (Edwards, 1972) to form DYNAMIX. A two-step solution approach is followed. As the numerical solution steps through time, the transport model is solved by explicit finite differences. Then, the chemical submodel computes the chemical distribution under dynamic partial equilibrium conditions. Liu and Narasimhan (1989b) verify the model by comparison with THCC and PHASEQL/FLOW (Walsh et al., 1984). They provide two example applications of the model, both of which involve numerous chemical components. A solution was provided to a two-dimensional contamination problem involving selenium and arsenic but including also 16 aqueous components, 26 minerals, and over 76 aqueous complexes. They found that arsenic was strongly retarded by precipitation as As_2S_3. On the other hand, selenium was relatively more mobile, being transported in the aqueous phase. Not surprisingly, a Cray XMP/14 supercomputer was used to perform the calculations.

3.6.2.3 FASTCHEM

The two-dimensional FASTCHEM (*F*ly *A*sh and Flue Gas Desulfurization *S*ludge *T*ransport and Geo*chem*istry) package consists of separate modules including a two-part thermodynamic database (Krupka et al., 1988), groundwater flow code (Huyakorn et al., 1988), streamtube code (Kincaid, 1988), and equilibrium geochemical code (Morrey, 1988). The separate modules can be run independently if desired. Other codes are used to link the modules and solve the coupled transport and geochemical transformation equations.

3.6.2.4 FIESTA

This one-dimensional model combines the FEAP and MINEQL models (Kincaid et al., 1984a). It will handle six components with reactions including ion exchange, adsorption, and aqueous phase complexation (Theis et al., 1982). A detailed evaluation of FIESTA was performed by Morrey et al. (1986), who were interested in disposal problems faced by the electricity industry. They found that the construction of the code was sound, but concluded that its capabilities could be improved substan-

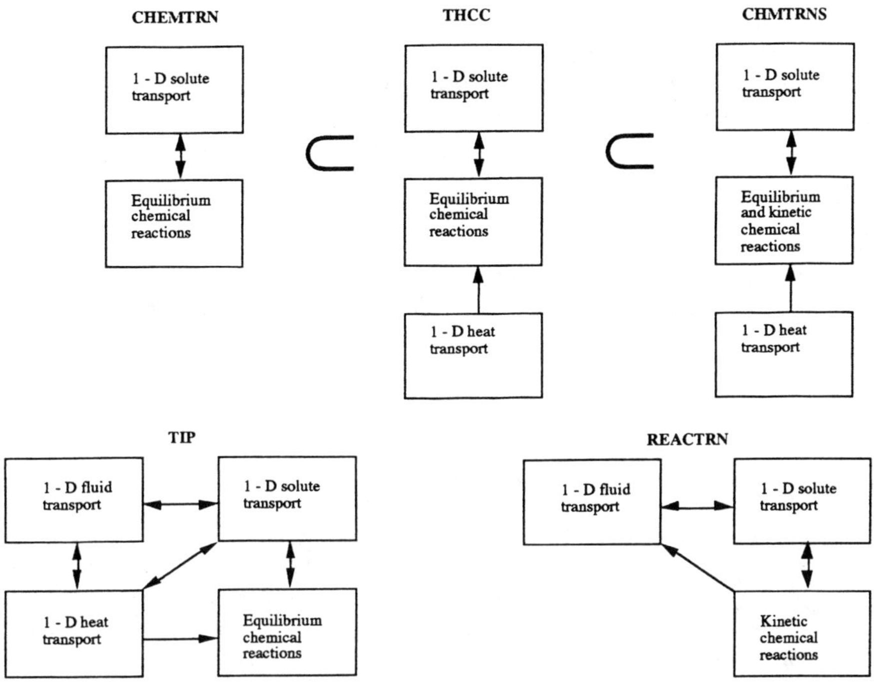

FIGURE 3.3. Processes modeled in various coupled flow, transport, and chemical reaction codes (cf. Carnahan, 1987). In the upper row the development of CHEMTRN is shown. In each box the processes modeled by the code are summarized. The arrows linking the boxes show the interaction between the various processes.

tially by, for example, increasing the number of components allowable, extending the adsorption, and allowing for variable dispersivity and velocity.

3.6.2.5 PHREEQM

PHREEQM combines the geochemical model PHREEQE with a one-dimensional mixing cell concept (Dance and Reardon, 1983; Schulz and Reardon, 1983), the latter to model the transport and dispersion of the solutes. Appelo and Willemsen (1987) performed a similar operation using the EQ3/6 (Wolery, 1983) geochemical model. In concept, the mixing-cell approach divides the domain into a group of cells. At each time step the fluid contained within each cell is advected and mixed between neighboring cells, with the mixing proportions between cells selected so that solute dispersion is simulated. Appelo et al. (1990) have used the model to simulate salt water transport and reactions in laboratory and field settings.

3.6.2.6 SATRA-CHEM

The flow and transport code, SATRA, a modified version of the SUTRA program (Voss, 1984), was combined with a mathematical formulation of the chemical reaction processes of sorption, ion exchange, and aqueous complexation to produce SATRA-CHEM (Lewis et al., 1987). These authors try explicitly to separate the linear and nonlinear portions of the numerical calculations. Linear governing equations can then be solved separately from the nonlinear equations with, it is hoped, a consequent gain in efficiency. The model solved by Lewis et al. (1987) is a relatively simple one, and no general procedure of separating the numerical tasks is presented. Numerous one- and two-dimensional examples of SATRA-CHEM are presented, but no experimental data are analyzed.

3.6.2.7 TIP

This program is based on the thermodynamics of irreversible processes. It solves the coupled equations of flow of heat, fluid, and solute for spatially varying temperature, pressure, and solute concentration. The original code was modified (Jacobsen and Carnahan, 1988) to include the reactive chemical modeling contained in the THCC program (Figure 3.3).

3.7 SEMIANALYTICAL TRANSPORT MODELS

3.7.1 A List of Major Models

Unlike the relatively few multicomponent transport models discussed above, there exists a plethora of fully analytical or semianalytical models that can be used to simulate the transport of a single chemical species (usually) or, rarely, more than one species. The many physical and chemical scenarios encompassed by models in this class are too numerous to attempt more than an overview of the literature and computer codes in this area. Extensive summaries have been prepared by Grove and Stollenwerk (1987) and Miller (1987), who review the characteristics of many recent publications. Some of the more widely used and recent models implemented on computers will be considered here. These codes, for the most part, are based on relatively simple spatial domains and chemical reaction models. Their solution usually involves either calculation of an analytical solution or some simple numerical scheme, e.g., the numerical inversion of a Laplace domain solution. A brief summary of the available codes is given in Table 3.4. More detailed descriptions of the codes and their applications are presented below.

3.7.2 Descriptions of Some Semianalytical Codes

3.7.2.1 COLUMN2

COLUMN2 is an extension of the COLUMN program, developed by Bo (1978). The latest version of the code couples an explicit finite-difference solution to the one-dimensional convection-dispersion equation equilibrium sorption, aqueous phase reactions, and radioactive decay. The nonlinearity of the chemical reactions is ignored by using previously calculated concentrations for the components in the calculations. Nielsen et al. (1985) provide a number of examples involving the use of COLUMN2, but no examples were compared with actual experimental data.

3.7.2.2 CXTFIT

CXTFIT consists of a suite of one-dimensional transport models to which data can be fitted to determine parameters. Concentration profiles or breakthrough curves can be produced if parameters are known. Any of the following models can be selected.

Linear Equilibrium Adsorption Models

$$R\frac{\partial c}{\partial t} = D\frac{\partial^2 c}{\partial x^2} - v\frac{\partial c}{\partial x} - \mu c + \lambda \tag{26}$$

Equation (26) is just the familiar convection-dispersion equation with first-order decay and zero-order production terms. It applies whether resident or flux concentrations (Barry and Sposito, 1988; Kreft and Zuber, 1978, 1986; Parker and van Genuchten, 1984b, 1986; Sposito and Barry, 1987) are considered.

Two-Site/Two-Region Nonequilibrium Models. This class of model considers "two-site" adsorption to be composed of equilibrium adsorption and first-order kinetic nonequilibrium adsorption (Selim et al., 1976; Rao et al., 1979). These adsorption sites are referred to as type-1 and type-2 sites, respectively. The governing model is

$$\left(1 + \frac{F\rho_b}{\theta}\right)\frac{\partial c}{\partial t} + \frac{\rho_b}{\theta}\frac{\partial s_2}{\partial t} = D\frac{\partial^2 c}{\partial x^2} - v\frac{\partial c}{\partial x} \tag{27a}$$

where

$$\frac{\partial s_2}{\partial t} = \alpha\left[(1-F)kc - s_2\right] \tag{27b}$$

Equation (27b) defines the model for the nonequilibrium, type-2 sites, with concentration s_2. The constant k is defined by the equilibrium relationship between the adsorbed and aqueous phase concentration:

$$s_1 + s_2 = kc \tag{28}$$

The "two-region" model is mathematically equivalent to the two-site model although it is based on a physical interpretation of the transport process, in contrast to the chemical adsorption model upon which the two-site model rests (van Genuchten, 1981). In the physical interpretation, the soil is subdivided conceptually into two regions, one labeled mobile and the other immobile. Flow takes place in the mobile region, but not in the immobile zone. However, solute moves into the immobile zone at a rate proportional to the local difference in concentration in each zone. Thus, even if experimental data are fitted well by the two-site or two-region model, additional information is needed to determine which interpretation is applicable.

Regional Transport Model. Many different models of field-scale transport models have been developed (e.g., Bresler and Dagan, 1983; Jury, 1982; Persaud et al., 1985; van der Zee and van Riemsdijk, 1986; van der Zee, 1990). Parker and van Genuchten (1984b) assume the field can be subdivided into one-dimensional, inde-

Table 3.4 Semianalytical Chemical Transport Models

Name	Description	Ref.
COLUMN2	One-dimensional transport code incorporating dispersion, sorption, ion exchange, and first and second order homogeneous chemical reactions	Nielsen et al. (1985)
CXTFIT	Code used to fit transport parameters for a number of one-dimensional models	Parker and van Genuchten (1984a)
MIGRATE	Quasi-two-dimensional transport with a linear adsorption isotherm	Rowe and Booker (1985)
MYGRT	Two-dimensional transport under steady flow conditions with retardation and linear decay	Summers et al. (1989)
PESTAN	One-dimensional convective transport with various simplified chemical processes	Enfield et al. (1982)
POLLUTE	Solves the one-dimensional solute transport equation for layered profiles	Rowe et al. (1984)
PRZM	This is composed of a hydrology component (runoff, erosion, evapotranspiration) and a chemical transport component (chemical uptake, decay, transformation, retardation)	Carsel et al. (1985)
SESOIL	A seasonal compartmental model of water flow, sediment, and solute transport applicable at the watershed scale	Bonazountas and Wagner (1984)
TFMFIT	Fits transport and chemical parameters for an areally averaged three-dimensional convection-dispersion model.	Barry et al. (1988)
TRANSS	A simple model for transport of a decaying radionuclide along a collection of streamtubes	Simmons et al. (1986)

pendent, parallel soil columns, with the solute transport in each governed by a convection-dispersion equation with its own transport parameters. The distribution of pore water velocity is described by the log-normal distribution:

$$f(v) = \left(v\sigma_{\ln}\right)^{-1}(2\pi)^{-1/2}\exp\left\{-\frac{\left[\ln(v)-\mu_{\ln}\right]^2}{2\sigma_{\ln}^2}\right\} \tag{29}$$

The distribution of the dispersion coefficient is taken as being uniform because transport effects due to this mechanism are small in comparison to velocity variability (Amoozegar-Fard et al., 1982). Under these assumptions, the mean field-scale resident solute concentration is calculated from

$$\bar{c} = \int_0^\infty c f(v)\, dv \tag{30}$$

The inlet condition, which is the local flux density at the inlet of each column, is

$$M(t) = \begin{cases} v\theta c_0 t, & 0 < t \le t_0 \\ v\theta c_0 t_0, & t > t_0 \end{cases} \tag{31}$$

Define M_0 by $M_0 = v\theta c_0 t_0$, so that M_0 is the total mass applied per unit area. Solutions are provided for the following combinations:

1. Sure t_0, θ, c_0; stochastic M_0, v
2. Sure M_0, θ, c_0; stochastic t_0, v

The fitting algorithm of CXTFIT is that of Marquardt (1963). Examples showing the uses of CXTFIT include fitting to various models (1) laboratory column data of Parker (1984) and van Genuchten (1974), and (2) the field data of Jury (1982). Barry and Parker (1987) extended CXTFIT to account for layered profiles. They obtained accurate predictions of breakthrough curves for experimental data presented by Shamir and Harleman (1967) and Panigatti (1970).

3.7.2.3 MIGRATE

This model is intended to be applied for contamination problems involving leachate emanating from a landfill. A quasi-two-dimensional vertical slice of the landfill is modeled. It is assumed that an initial, known quantity of leachate is stored in the landfill at a given concentration. It moves downward at a steady velocity through a clay layer until it encounters a more permeable layer in which the flow is horizontal. Only linear equilibrium solute adsorption is included in the model. The Laplace domain solution derived by Rowe and Booker (1985b) is solved using the numerical inversion routine of Talbot (1979). Hensley (1989) and Hensley and Schofield (1991) compared MIGRATE with experimental data from a centrifuge experiment simulating a landfill leachate plume, and found that it was not able to replicate the experimental results at all points in the centrifuge. In particular, MIGRATE predicted poorly transverse solute transport.

3.7.2.4 MYGRT

This code is based on the two-dimensional convection-dispersion equation with a linear decay term on a semiinfinite domain. A one-dimensional version of the same governing equation can be selected as well. In both cases analytical solutions are available (Cleary and Ungs, 1978; Javandel et al., 1984). The main computations performed are integrations that are carried out numerically. Solutions for an impermeable base at the lower boundary of the profile are obtained using the method of images.

3.7.2.5 PESTAN

This is a one-dimensional model that accounts for aqueous-phase transport and transformations under steady flow in homogeneous soil profiles. Three solutions are provided. These are (1) linear sorption and first-order decay without dispersion, (2) dispersion and linear sorption without decay, and (3) adsorption modeled using the Freundlich isotherm with first-order decay but no dispersion. Cases (1) and (2) are solved analytically while case (3) is solved numerically. PESTAN has been used by the U.S. Environmental Protection Agency Office of Pesticide Programs (Donigian and Rao, 1986a) and by Jones and Back (1984) to model aldicarb transport in a Florida soil. Donigian and Rao (1986b) compared the predictions of PESTAN with aldicarb data presented by Hornsby et al. (1983). They found reasonable agreement for leaching predictions in this case.

3.7.2.6 POLLUTE

This code is essentially a one-dimensional version of MIGRATE. It simulates one-dimensional vertical flow from a finite source of solute with linear adsorption as the only chemical reaction considered (Rowe and Booker, 1985a). Below this finite (clayey) layer is another, more permeable layer. The more permeable layer is perfectly mixed in the vertical direction and ignores dispersion in the horizontal direction. The lower boundary condition is obtained by conservation of mass of solute entering the permeable layer. The experiments described by Hensley and Schofield (1991) produced a flow field that was essentially unidimensional for at least part of the flow domain. Data obtained from this subdomain were predicted well using POLLUTE.

3.7.2.7 PRZM

Flow in PRZM is modeled using an empirical model (Donigian and Rao, 1986b). PRZM is a compartmental model, with the chemical submodel including plant uptake of solute, transport by surface runoff and erosion, solute decay, retardation, and transformation, as well as transport by dispersion. The profile is divided into layers, with a one-dimensional convection-dispersion model applicable in each layer. In the surface layer, the model accounts for erosion and runoff-induced chemical transport. A numerical solution is obtained using the finite-difference method. It has been used for the modeling of pesticide transport in the analysis of several field studies (Jones et al., 1983; Jones, 1983; Carsel et al., 1985, 1986).

3.7.2.8 SESOIL

SESOIL is a compartmental model aimed at predicting solute distribution in both the soil profile and watershed on a seasonal basis. The code has a hydrologic cycle component based on the theory of Eagelson (1978). Both sediment and chemical transport are modeled by SESOIL. Chemical transformations and reactions included are volatilization, sorption, degradation, hydrolysis, oxidation, and cation exchange, with extensions to include other chemical processes planned (Donigian and Rao, 1986a). Donigian and Rao (1986a) report on a comprehensive evaluation of SESOIL. It was concluded that several improvements could be made to the chemical transport part of the code. Large variations in the hydraulic properties of the profile could not be handled well. The estimate of the time necessary for a chemical to leach to groundwater did not include sorption of the chemical, only the water movement in the profile. Donigian and Rao (1986b) used SESOIL to predict aldicarb movement to groundwater. They found that SESOIL overpredicted the mass of the pesticide leached. They also noted an error in the hydrolysis algorithm in the model.

3.7.2.9 TFMFIT

TFMFIT is based on the following convection-dispersion model (cf. Sposito and Jury, 1988):

$$BR\frac{\partial c_{\mathrm{m}}}{\partial t}+(1-B)\frac{\partial c_{\mathrm{m}}}{\partial t}=\sum_{i=1}^{3}\frac{1}{P_i}\frac{\partial^2 c_{\mathrm{m}}}{\partial X_i^2}-\frac{\partial c_{\mathrm{m}}}{\partial X_3}-BCc_{\mathrm{m}}-(1-B)Dc_{\mathrm{im}} \tag{32a}$$

with

$$(1-B)R\frac{\partial c_{\mathrm{im}}}{\partial t}+(1-B)Dc_{\mathrm{im}}=W\left(c_{\mathrm{m}}-c_{\mathrm{im}}\right) \tag{32b}$$

where

$$B=\frac{\theta_{\mathrm{m}}+f\rho_{\mathrm{b}}K_{\mathrm{d}}+\theta_{\mathrm{gm}}K_{\mathrm{H}}^{-1}}{\theta+\rho_{\mathrm{b}}K_{\mathrm{d}}+\theta_{\mathrm{g}}K_{\mathrm{H}}^{-1}} \tag{32c}$$

$$C=\frac{\left(\theta_{\mathrm{m}}\beta_{\mathrm{m}}^{1}+fK_{\mathrm{d}}\beta_{\mathrm{m}}^{\mathrm{s}}+\theta_{\mathrm{gm}}K_{\mathrm{H}}^{-1}\beta_{\mathrm{im}}^{\mathrm{g}}\right)L}{\theta_{\mathrm{m}}v(1-B)} \tag{32d}$$

$$D = \frac{\left(\theta_{\text{im}}\beta^{\text{l}}_{\text{im}} + (1-f)K_{\text{d}}\beta^{\text{s}}_{\text{im}} + \theta_{\text{im}} + \theta_{\text{gim}}K_{\text{H}}^{-1}\beta^{\text{g}}_{\text{im}}\right)L}{\theta_{\text{m}}v(1-B)} \tag{32e}$$

and

$$W = \frac{\left(\alpha^{\text{l}} + \alpha^{\text{g}}K_{\text{H}}^{-1}\right)L}{\theta_{\text{m}}v} \tag{32f}$$

This model represents the following processes (Barry et al., 1988):

1. Solute is free to move in any spatial direction through a porous medium with uniform porosity, with steady flow along the X_3 axis.
2. Solute dispersion is anisotropic.
3. Solute can be partitioned between the gas phase, the mobile and immobile portions of the liquid phase, and the solid phase, with biodegradation possible in all phases.

Barry et al. (1988) provided both real-time and Laplace domain solutions for Equation (29) averaged over a window that corresponds, approximately, to the sampling region of an experiment. TFMFIT can fit experimental data or produce concentration profiles or breakthrough curves for specified parameter sets. Like CXTFIT, TFMFIT uses the Marquardt (1963) algorithm to carry out the fitting portion of the computations. The user can choose to use either the numerically inverted Laplace domain solution or the real time solution.

TFMFIT was used to fit carbon tetrachloride breakthrough curve data from the Borden aquifer experiment (Roberts et al., 1986). As shown in Figure 3.4, the model fitted to the data very well, and found, for example, that a retardation coefficient of 1.74 was appropriate. TFMFIT assumes a Dirac source term, so it computes the probability density function of the solute transport to a particular location. Finite duration, non-Dirac source terms are permissible so long as the distance from the source is relatively large. Tetrachloroethylene breakthrough curve data for the same experiment, described by Curtis et al. (1986), were also fitted. Again a good model fit was found (Figure 3.5), and it was shown conclusively that, in terms of the model used, tetrachloroethylene is more strongly retarded (R = 2.47) than carbon tetrachloride.

3.7.2.10 TRANSS

The basic concept behind TRANSS bears some similarity to the regional transport model contained in CXTFIT. The transport domain is conceptualized as a collection of streamtubes, each with different transport properties. A one-dimensional convection-dispersion equation is assumed to apply in each streamtube. The final concentration through the exit surface is a probability-weighted average over all the streamlines, with the weighting predetermined by the user. The model can treat radioactive or conservative solutes, with decay in both the source and streamtube. The only

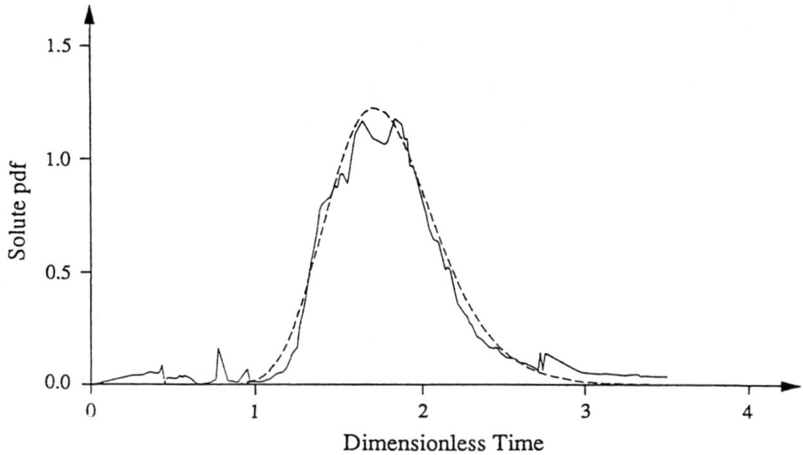

FIGURE 3.4. Fit of the convection-dispersion model, Equation (32) (---), to the carbon tetrachloride breakthrough curve data (—) from the Borden aquifer experiment, as performed by TFMFIT.

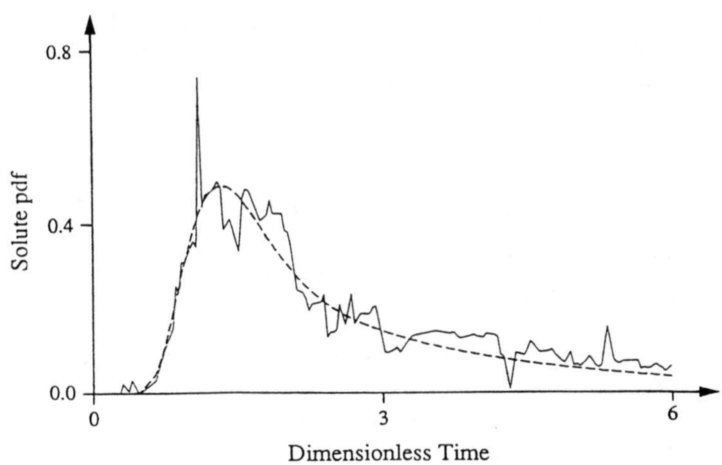

FIGURE 3.5. Fit of the convection-dispersion model, Equation (32) (---) to the tetrachloroethylene data (—) collected from the Borden aquifer experiment as perfored by TFMFIT.

chemical reaction included is retardation due to solute sorption. Input concentrations can be stochastic or sure. Solute release from the source can be modeled in four different ways: (1) constant release rate, (2) solubility-controlled release, (3) adsorption-equilibrium-controlled release, and (4) diffusion-controlled release from below an infiltration barrier.

3.8 SUMMARY

The emphasis of many hydrologic modelers has been on the physical processes of convection and dispersion of solutes. At the same time, many batch-type geochemical models have been implemented. It is easy to imagine situations requiring both aspects of contaminant transport to be incorporated in a single model. Thus, this chapter began with the premise that it is necessary to quantify both chemical and physical processes in contaminant transport models.

The basic structure of the geochemical models was outlined. This structure can be incorporated into a flow and transport model, yielding a set of coupled nonlinear partial differential equations to be solved. The number of computer codes that have been developed for the multicomponent transport model is small. Even fewer models have been developed for multidimensional spatial domains. The reason is that any problem involving a significant number of chemical reactions will require the availability of a powerful computer. In some cases a supercomputer will be necessary. In cases where it is thought that only a single or a few chemical components are involved, numerical solutions can be obtained with a much more modest computer.

3.9 LIST OF SYMBOLS

$a_{i,j}$	i = 1, … , *N*; *j* = 1, … , *M*, stoichiometric coefficient of the *j*th component in the *i*th species
A	Empirical constant in the activity coefficient formula
A_{ij}	Stoichiometric coefficient
B	Empirical constant in the activity coefficient formula
B	Dimensionless constant
c, c_i	$i = 1, \ldots, N$; species molarity, $M\,L^{-3}$
c_i^{T}	Total concentration of the *i*th component, $M\,L^{-3}$
c_{im}	Concentration in the immobile region, $M\,L^{-3}$
c_{m}	Concentration in the mobile region, $M\,L^{-3}$
c_0	Solute concentration of the influent fluid, $M\,L^{-3}$
C	Dimensionless parameter
D	Hydrodynamic dispersion coefficient, $L^2\,T^{-1}$
D	Hydrodynamic dispersion tensor, $L^2\,T^{-1}$
E_{F}	Equivalent fraction of *F* in the solid phase
$\tilde{E}_{\mathrm{F}}$	Equivalent fraction of *F* in the liquid phase
f	Probability density function
f	Fraction of adsorbing solid phase in contact with mobile liquid
f^{cu}	Nonlinear function of u_i, $i = 1, \ldots, N_{\mathrm{c}}$, $M\,L^{-3}$
f^{su}	Generalized adsorption isotherm, $M\,L^{-3}$
F	Exchange cation with valence *f*
F	Fraction of type-1 sorption sites
G	Exchange cation with valence *g*
k	Proportionality constant between adsorbed and fluid phase concentration

k_i	$i = 1, 4$; adjustable parameters
k_i	$i = 2, 5$; adjustable parameters, $L^3 M^{-1}$
k_3	Adjustable parameter, $L^{3(k_1-1)}M^{1-k_1}$
k_6	Adjustable parameter, $L^{3/2}M^{-1/2}$
k_7	Adjustable parameter, ML^{-3}
k_8	Adjustable parameter, $M^{k_1}L^{-3k_1}$
k_9	Adjustable parameter, $L^{3(1-k_1)}M^{k_1-1}$
k_{10}	Adjustable parameter, $L^{3(c^{k_1}/k_8-1)}M^{1-c^{k_1}/k_8}$
K_d	Distribution coefficient for adsorption, $L^3 M^{-1}$
K_e, K_i, $K_{e,i}$	$i = 1, \ldots, N$; equilibrium constants
K_H	Henry's law constant for solute partitioning between the liquid and air
K_{s_i}	Equilibrium constant for adsorption reaction
K_{sp}	Solubility product
$\mathbf{L}$	Convection-dispersion differential operator, T^{-1}
L	Distance form solute input point to monitoring point, L
m, m_i, etc.	Molality
m_{sol}	Mass of solvent, M
m_{sp}	Mass of a chemical species, M
$M(t)$	Cumulative flux of solute at time t, $M L^{-2}$
M_0	Total amount of solute added to profile, $M L^{-2}$
M_{Tw}	Mass of water, M
N_c	Number of chemical components
N_x	Number of chemical complexes
P_i	Peclet number in the ith direction
q	Excess of chemical adsorbed
$\mathbf{q}$	Fluid flux, $L T^{-1}$
r_i	$i = 1, \ldots, N$, stoichiometric coefficients
R	Retardation coefficient
R_i	$i = 1, \ldots, N$; reacting chemical species
s	Concentration of adsorbed solute, $M L^{-3}$
s_1, s_2	Solute concentration on type-1 and -2 sites, respectively, $M L^{-3}$
s_i	$i = 1, \ldots, M$; stoichiometric coefficients
S_i	$i = 1, \ldots, M$; product of reacting chemical species
t	Dimensionless time coordinate
t_i	$i = 1, \ldots, N$; stoichiometric coefficients
TN	Total normality of a solution
u_i	Total aqueous phase concentration of the ith component, $M L^{-3}$
v	Mean pore water velocity, $L T^{-1}$
V_{sol}	Volume of solution, L^3
w	Sorbed concentration, $M L^{-3}$
x_i	Concentration of the ith complex, $M L^{-3}$
x_T	Total number of surface sites in moles per unit volume of solution
X	Anion with valence –1
X_j	$j = 1, \ldots, M$; concentration of the jth component, $M L^{-3}$

X_i	Dimensionless distance in the ith direction
$X_{T,j}$	$j = 1, \ldots, M$; total concentration of the jth component in the system, $M\,L^{-3}$
Y	Anion with valence −1
z, z_i, etc.	Valence
$Z^{\mp z_Z}$	Exchange ion with valence z_Z
α	First-order rate coefficient, T^{-1}
α^i	Linear solute transfer coefficient for the ith (gas, g; liquid, l) component into the immobile portion of the liquid phase, T^{-1}
β^i_j	First-order solute decay constant for the ith (gas, g; liquid, l; solid, s) component in or contacting the jth (mobile, m; immobile, im) portion of the liquid phase, T^{-1}
γ	Activity coefficient, $L^3\,M^{-1}$
θ	Volumetric moisture content
θ_{im}	Volumetric moisture content of the immobile region
θ_{gim}	Volumetric air content in contact with the immobile region
θ_{gm}	Volumetric air content in contact with the mobile region
θ_m	Volumetric moisture content of the mobile region
λ	Zero-order production coefficient, $M\,L^{-3}\,T^{-1}$
μ	Rate constant for first-order decay, T^{-1}
μ_{ln}	Mean of $\ln(v)$
ρ_b	Bulk density, $M\,L^{-3}$
σ_{ln}	Standard deviation of $\ln(v)$
$\{\bullet\}$	Activity of •
$\bullet_l$	In the liquid phase
$\bullet_s$	In the solid phase
$\bar{\bullet}$	Mean of •

3.10 APPENDIX: CHEMISTRY CONCEPTS

The term *chemical species* may refer to either single ionic species or chemical *complexes,* i.e., ionic species acting as a single unit. The concentration of a solute can be expressed in various ways. Perhaps the most general, because it is independent of temperature or pressure changes, is the *molality,* which is simply the ratio of the mass of the chemical species to the mass of the solvent (Sposito, 1989, p. 264), i.e.,

$$m = \frac{m_{sp}}{m_{sol}} \tag{A1}$$

Molality is usually measured in units of moles per kilogram of solvent. More commonly, the molarity is used as the concentration measurement. *Molarity* is defined by (Sposito, 1981, §2.2):

$$c = \frac{m_{\mathrm{sp}}}{V_{\mathrm{sol}}} \tag{A2}$$

and is usually expressed as moles per cubic meter. The concentration of a solute does not necessarily define its capacity to react. This is because of interaction between solute particles themselves or interactions with the solvent. The *activity* is a measure of the effective concentration of a solute (Tan, 1982, §3.15). If a solute has concentration c, then its activity is given by

$$\{c\} \equiv \gamma c \tag{A3}$$

The *activity coefficient*, γ, has units of liters per mole, so that the activity is dimensionless (Sposito, 1989, §4.5). One can also define single ion activity coefficients by analogy with Equation (A3), although Sposito (1984b) points out that the meaning of such coefficients is purely operational. Activity is a measure of the chemical potential of the solute (Schay, 1969, §B). The activity coefficient is less than unity, although it approaches unity for dilute solutions. Thus, because many contaminant problems concern dilute solutes, the concentration of a species will be a good approximation to its activity. When the species is adsorbed onto the solid phase, the concentration variable changes to mole fractions. Mole fraction is defined as the ratio of the number of moles of the species under consideration to the total number of moles of adsorbed species. In this case the activity coefficient is termed a *rational activity coefficient* (Sposito, 1979, §9.4). In multicomponent solutions of electrolytes, the activity coefficient can be calculated using the equation

$$\log(\gamma) = BI - \frac{Az^2 I\sqrt{\mathrm{I}}}{1 + I\sqrt{\mathrm{I}}} \tag{A4}$$

where the *ionic strength*, I, of the solution is given by

$$I = \frac{1}{2}\sum_{i=1}^{n_s} z_i^2 m_i \tag{A5}$$

There are various versions of Equation (A4) available, depending on the (temperature-dependent) values of A and B (see Sposito, 1981, §2.6 for a range of values). Davies (1962) has produced an expression that is accurate for ionic strengths less than 0.5 mol/kg (Sposito, 1984b). If a reaction occurs in an aqueous solution without precipitation or adsorption, then at equilibrium the *law of mass action* asserts that the ratio of activities of the products to reactants is constant. The factor of proportionality is the *equilibrium constant* (Tan, 1982, §3.5). Ions that become associated in a reaction form a *complex* (Sposito, 1989, §4.2). Suppose the species $R_1, \ldots, R_N$ react to form products $S_1, \ldots, S_M$, according to

$$r_1R_1 + \cdots + r_NR_N \rightleftarrows t_1S_1 + \cdots + t_MS_M \tag{A6}$$

then the equilibrium constant, K_e, which is temperature dependent, is given by

$$K_e = \frac{\{R_1\}^{r_1} \times \cdots \times \{R_N\}^{r_N}}{\{S_1\}^{t_1} \times \cdots \times \{S_M\}^{t_M}} \tag{A7}$$

If we make use of Equation (A3), then Equation (A7) may be rewritten as

$$K_e = \frac{\gamma_{R_1}^{r_1} R_1^{r_1} \times \cdots \times \gamma_{R_N}^{r_N} R_N^{r_N}}{\gamma_{S_1}^{t_1} S_1^{t_1} \times \cdots \times \gamma_{S_M}^{t_M} S_M^{t_M}} \tag{A8}$$

3.11 REFERENCES

Amoozegar-Fard, A., D. R. Nielsen, and A. W. Warrick. 1982. Soil solute concentration distributions for spatially varying pore water velocities and apparent diffusion coefficients. *Soil Sci. Soc. Am. J.* 46:3–9.

Appelo, C. A. J. and A. Willemsen. 1987. Geochemical calculations and observations on salt water intrusions. I. A combined geochemical/mixing cell model. *J. Hydrol.* 94:313–330.

Appelo, C. A. J., A. Willemsen, H. E. Beekman, and J. Griffioen. 1990. Geochemical calculations and observations on salt water intrusions. II. Validation of a geochemical model with laboratory experiments. *J. Hydrol.* 120:225–250.

Baes, C. F. and R. E. Mesmer. 1976. The Hydrolysis of Cations. John Wiley & Sons, New York.

Ball, J. W., E. A. Jenne, and M. W. Cantrell. 1981a. WATEQ3: A Geochemical Model with Uranium Added. U.S. Geological Survey, Open-File Report 81-1183, Menlo Park, CA.

Ball, J. W., D. K. Nordstrom, and E. A. Jenne. 1981b. Additional and Revised Thermochemical Data and Computer Code for WATEQ2 — A Computerized Chemical Model for Trace and Major Element Speciation and Mineral Equilibria of Natural Waters. U.S. Geological Survey Water Resources Investigations WRI 78-116 (Second Printing), Menlo Park, CA.

Bajpai, A. C., L. R. Mustoe, and D. Walker. 1977. Advanced Engineering Mathematics. John Wiley & Sons, Chichester, England.

Barragan, R. M. and D. Nieva. 1989. EQQYAC: program for determining geothermal reservoir chemical equilibrium. *Comput. Geosci.* 15(8):1221–1240.

Barrer, R. M. and J. Klinowski. 1974. Ion-exchange selectivity and electrolyte concentration. *J. Chem. Soc. Faraday Trans. 1* 70:2080–2091.

Barrer, R. M. and J. Klinowski. 1979. Cation exchangers with several groups of sites. *J. Chem. Soc. Faraday Trans. 1* 75:247–251.

Barrow, N. J. 1989. Suitability of sorption-desorption models to simulate partitioning and movement of ions in soils. In: Inorganic Contaminants in the Vadose Zone. B. Bar-Yosef, N. J. Barrow, and J. Goldsmith (Eds.) Springer-Verlag, Berlin, pp. 18–42.

Barry, D. A. 1990a. Comments on "Estimating Michaelis-Menten or Langmuir isotherm constants by weighted nonlinear least squares" by P. Persoff and J. F. Thomas. *Soil Sci. Soc. Am. J.* 54(3):941–942.

Barry, D. A. 1990b. Supercomputers and their use in modeling subsurface solute transport. *Rev. Geophys.* 28(3):277–295.

Barry, D. A. and J. C. Parker. 1987. Approximations for solute transport through porous media with flow transverse to layering. *Transp. Porous Media* 2:65–82.

Barry, D. A. and G. Sposito. 1988. Application of the convection-dispersion model to solute transport in finite soil columns. *Soil Sci. Soc. Am. J.* 52(1):3–9.

Barry, D. A., G. Sposito, and J. Coves. 1988. TFMFIT: A Program to Fit a Generalized Convection-Dispersion Model to Experimental Data. Department of Soil and Environmental Science, University of California, Riverside, CA.

Bear, J. 1972. Dynamics of Fluids in Porous Media. Elsevier, New York.

Bo, P. 1978. COLUMN — Numerical Solution of Migration Equations Involving Various Physicochemical Processes. Chemistry Department, Risø National Laboratory, Roskilde, Denmark.

Bonazountas, M. and J. Wagner. 1984. SESOIL: A Seasonal Soil Compartment Model. Arthur D. Little, Inc., Cambridge, MA.

Bowers, T. S., K. L. Von Damm, and J. M. Edmond. 1985. Evolution of mid-ocean ridge hot springs. *Geochim. Cosmochim. Acta* 49(11):2239–2252.

Bresler, E. and G. Dagan. 1983. Unsaturated flow in spatially variable fields. 3. Solute transport models and their application to two fields. *Water Resour. Res.* 19:429–435.

Brinkley, S. R. 1946. Note on the conditions of equilibrium for systems of many constituents. *J. Chem. Phys.* 14(9):563–564.

Brinkley, S. R. 1947. Calculation of equilibruim composition of systims of many constituents. *J. Chem. Phys.* 15(2):107–110.

Brown, A. L. and A. Page. 1970. Elements of Functional Analysis. Van Nostrand Reinhold, New York.

Bryant, S. L., R. S. Schechter, and L. W. Lake. 1986. Interactions of precipitation/dissolution waves and ion exchange in flow through permeable media. *Am. Inst. Chem. Eng. J.* 32(5):751–764.

Butt, J. B. 1980. Reaction Kinetics and Reactor Design. Prentice-Hall, Englewood Cliffs, NJ.

Carnahan, C. L. 1987a. Effects of coupled thermal, hydrological and chemical processes on nuclide transport. Paper presented at GEOVAL-87, a symposium sponsored by the Swedish Nuclear Power Inspectorate, Stockholm, Sweden, April 7–9.

Carnahan, C. L. 1987b. Simulation of chemically reactive solute transport under conditions of changing temperature. In: Coupled Processes Associated with Nuclear Waste Repositories. C.-F. Tsang (Ed.). Academic Press, Orlando, FL, pp. 249–257.

Carnahan, C. L. and J. S. Jacobsen. 1988. Coupled Transport Processes in Semipermeable Media. Part I. Theoretical Basis. Lawrence Berkeley Laboratory Report LBL-25618, University of California, Berkeley, CA.

Carsel, R. F., L. A. Mulkey, M. N. Lorber, and L. B. Baskin. 1985. The pesticide root zone model (PRZM): a procedure for evaluating pesticide leaching threats to groundwater. *Ecol. Model.* 30:49–69.

Carsel, R. F., W. B. Nixon, and L. G. Balentine. 1986. Comparison of pesticide root zone model predictions with observed concentrations for the tobacco pesticide metalaxyl in unsaturated zone soils. *Environ. Toxicol. Chem.* 5:345–353.

Cleary, R. W. and M. U. Ungs. 1978. Groundwater Pollution and Hydrology: Mathematical Models and Computer Programs. Report No. 78-WR-15. Water Resources Program, Princeton University, Princeton, NJ.

Coves, J. and G. Sposito. 1986. MICROQL-UCR: A Surface Chemical Adaption of the Speciation Program MICROQL. Department of Soil and Environmental Sciences, University of California, Riverside, CA.

Curtis, G. P., P. V. Roberts, and M. Reinhard. 1986. A natural gradient experiment on solute transport in a sand aquifer. 4. Sorption of organic solutes and its influence on mobility. *Water Resour. Res.* 22(13):2059–2967.

Dance, J. T. and E. J. Reardon. 1983. Migration of contaminants in groundwater at a landfill: a case study. 5. Cation migration in the dispersion test. *J. Hydrol.* 63:109–130.

Davies, C. W. 1962. Ion Association. Butterworths, London.

Deland, E. C. 1967. CHEMIST — The Rand Chemical Equilibrium Program. Memorandum RM-5404-PR.

Donigian, A. S. and P. S. C. Rao. 1986a. Overview of terrestrial processes and modeling. In: Vadose Zone Modeling of Organic Pollutants, Ch. 1. S. C. Hern and S. M. Melancon (Eds.). Lewis, Chelsea, MI.

Donigian, A. S. and P. S. C. Rao. 1986b. Example model testing studies. In: Vadose Zone Modeling of Organic Pollutants, Ch. 5. S. C. Hern and S. M. Melancon (Eds.). Lewis, Chelsea, MI.

Eagelson, P. S. 1978. Climate, soil and vegetation. *Water Resour. Res.* 14:1–7.

Edwards, A. L. 1973. TRUMP: A Computer Program for Transient and Steady State Temperature Distributions in Multi-Dimensional Systems. Report No. 14654, Review 3. Lawrence Livermore National Laboratory, Livermore, CA.

Enfield, C. G., R. F. Carsel, S. Z. Cohen, T. Phan, and D. M. Walters. 1982. Approximating pollutant transport to ground water. *Ground Water* 20:711–722.

Felmy, A. R., D. Girvin, and E. A. Jenne. 1984. MINTEQ: A Computer Program for Calculating Aqueous Geochemical Equilibria. U.S. Environmental Protection Agency, Athens, GA.

Felmy, A. R., A. E. Reisenauer, J. M. Zachara, and G. W. Gee. 1984. MININR: A Geochemical Computer Program for Inclusion in Water Flow Models — an Applications Study. Report No. PNL-4921. Pacific Northwest Laboratory, Richland, WA.

Fitter, A. H. and C. D. Sutton. 1975. The use of the Freundlich isotherm for soil phosphate sorption data. *J. Soil Sci.* 26:241–246.

Fournier, R. O. and R. W. Potter. 1982. A revised and expanded silica quartz geothermometer. *Geotherm. Resour. Counc. Bull.* 11(10):3–12.

Freundlich, H. 1909. Kapillarchemie; eine Darstellung der Chemie der kolloide und verwandter Gebiete. Akademische Verlagsgellschaft, Leipzig.

Freundlich, H. 1926. Colloid and Capillary Chemistry. Methuen, London.

Furrer, G., J. Westall, and P. Sollins. 1989. The study of soil chemistry through quasi-steady-state models. I. Mathematical definition of model. *Geochim. Cosmochim. Acta* 53:595–601.

Garrels, R. M. and M. E. Thompson. 1962. A chemical model for sea water at 25°C and one atmosphere total pressure. *Am. J. Sci.* 260(1):57–66.

Giggenbach, W. F. 1980. Geothermal gas equilibria. *Geochim. Cosmochim. Acta* 44(12):2021–2032.

Grove, D. B. and K. G. Stollenwerk. 1987. Chemical reactions simulated by ground-water-quality models. *Water Resour. Bull.* 23(4):601–615.

Gunary, D. 1970. A new adsorption isotherm for phosphate in soil. *J. Soil Sci.* 21:72–77.

Hassanizadeh, M. and W. G. Gray. 1979a. General conservation equations for multi-phase systems. 1. Averaging procedure. *Adv. Water Resour.* 2:131–144.

Hassanizadeh, M. and W. G. Gray. 1979b. General conservation equations for multiphase systems. 2. Mass, momenta, energy, and entropy equations. *Adv. Water Resour.* 2:191–203.

Helgeson, H. C., W. M. Murphy, and P. Aagaard. 1984. Thermodynamic and kinetic constraints on reaction rates among minerals and aqueous solutions. II. Rate constants, effective surface area, and the hydrolysis of feldspar. *Geochim. Cosmochim. Acta* 48:2405–2432.

Henley, R. W., A. H. Truesdell, P. B. Barton, and J. A. Whitney. 1984. Fluid-Mineral Equilibria in Hydrothermal Systems. *Rev. Econ. Geol.,* Vol. 1, Soc. Econ. Geol.

Hensley, P. J. 1989. Accelerated Physical Modelling of Transport Processes in Soil. Ph.D. thesis. Cambridge University, Cambridge, England.

Hensley, P. J. and A. N. Schofield. 1991. Accelerated physical modelling of hazardous waste transport. *Geotechnique* 41(3):447–466.

Holford, I. C. R. 1982. The comparative significance and utility of the Freundlich and Langmuir parameters for characterising sorption and plant availability of phosphate in soils. *Aust. J. Soil Res.* 20:233–242.

Hornsby, A. G., P. S. C. Rao, W. B. Wheeler, P. Nkedi-Kizza, and R. L. Jones. 1983. Fate of Aldicarb in Florida Citrus Soils. 1. Field and Laboratory Studies. Proc. of the NWWA/U.S. EPA Conference on Characterization and Monitoring of the Vadose (Unsaturated) Zone, Las Vegas, NV.

Hostetler, C. J., R. L. Erikson, J. S. Fruchter, and C. T. Kincaid. 1988. Overview of FASTCHEM™ Package: Application to Chemical Transport Problems. Report EA-5870, Vol. 1. Electric Power Research Institute, Palo Alto, CA.

Huyakorn, P. S., V. M. Guvanasen, T. D. Wadsworth, and B. H. Lester. 1988. FASTCHEM™ Package: User's Guide to the EFLOW Groundwater Flow Code. Report EA-5870-CCM, Vol. 2. Electric Power Research Institute, Palo Alto, CA.

Jacobsen, J. S. and C. L. Carnahan. 1988. Numerical simulation of alteration of sodium bentonite by diffusion of groundwater components. In: Scientific Basis for Nuclear Waste Management XI. *Mater. Res. Soc. Symp. Proc.* 112.

Javandel, I., C. Doughty, and C. Tsang. 1984. Groundwater Transport: Handbook of Groundwater Models. Water Resources Monograph 10. American Geophysical Union, Washington, D.C.

Jaycock, M. J. and G. D. Parfitt. 1981. Chemistry of Interfaces. Ellis Horwood, Chichester, England.

Jenne, E. A. 1981. Groundwater Modeling: A Review. Report No. PNL-3754. Pacific Northwest Laboratory, Richland, WA.

Jones, R. L. 1983. Movement and degradation of aldicarb residues in soil and groundwater. Paper presented at the Environmental Toxicology and Chemistry Conference on Multidisciplinary Approaches to Environmental Problems, Crystal City, VA.

Jones, R. L. and R. C. Back. 1984. Monitoring aldicarb residues in Florida soil and water. *Environ. Toxicol. Chem.* 3:9–20.

Jones, R. L., P. S. C. Rao, and A. G. Hornsby. 1983. Fate of aldicarb in Florida citrus soil. 2. Model evaluation. Proceedings of the NWWA/U.S. EPA Conference on Characterization and Monitoring of the Vadose (Unsaturated) Zone, Las Vegas, NV.

Jury, W. A. 1982. Simulation of solute transport using a transfer function model. *Water Resour. Res.* 18:363–368.

Kerrisk, J. 1984. Reaction-Path Calculations of Ground-Water Chemistry and Mineral Formation of Rainier Mesa, Nevada. LANL-9912-MS. Los Alamos National Laboratory, Los Alamos, NM.

Kharaka, I. K. and I. Barnes. 1973. SOLMNEQ: Solution Mineral Equilibrium Computations. Geological Survey Computer Control Publication #215899. U.S. Department of the Interior, Washington, D.C.

Kincaid, C. T. 1988. FASTCHEM™ Package: User's Guide to ETUBE, The Pathline and Streamtube Database Code. Report EA-5870, Vol. 3. Electric Power Research Institute, Palo Alto, CA.

Kincaid, C. T., J. R. Morrey, and J. E. Rogers. 1984a. Geohydrochemical Models for Solute Migration: Process Description and Computer Code Selection, Vol. 1. Report EA-3417. Electric Power Research Institute, Palo Alto, CA.

Kincaid, C. T., J. R. Morrey, S. B. Yabusaki, A. R. Felmy, and J. E. Rogers. 1984b. Geohydrochemical models for solute migration: Preliminary Evaluation of Selected Computer Codes, Vol. 2. Report EA-3417. Electric Power Research Institute, Palo Alto, CA.

Kinniburgh, D. G. 1986. General purpose adsorption isotherms. *Environ. Sci. Technol.* 20(9):895–904.

Kinniburgh, D. G., J. A. Barker, and M. Whitfield. 1983. A comparison of some simple adsorption isotherms for describing divalent cation adsorption by ferrihydrite. *J. Colloid Interface Sci.* 95(2):370–384.

Kirkner, D. J. and H. W. Reeves. 1988. Multicomponent mass transport with homogeneous and heterogeneous chemical reactions: effect of the chemistry on the choice of numerical algorithm. 1. Theory. *Water Resour. Res.* 24(10):1719–1729.

Kirkner, D. J. and H. W. Reeves. 1990. A penalty function method for computing chemical equilibria. *Comput. Geosci.* 16(1):21–40.

Kirkner, D. J., T. L. Theis, and A. A. Jennings. 1984. Multicomponent solute transport with sorption and soluble complexation. *Adv. Water Resour.* 7:120–125.

Kreft, A. and A. Zuber. 1978. On the physical meaning of the dispersion equation and its solutions for different initial and boundary conditions. *Chem. Eng. Sci.* 33:1471–1480.

Kreft, A. and A. Zuber. 1986. Comments on "Flux-averaged and volume-averaged concentrations on continuum approaches to solute transport" by J. C. Parker and M. Th. van Genuchten. *Water Resour. Res.* 22(7):1157–1158.

Krupka, K. M., R. L. Erikson, S. V. Mattigod, J. A. Schramke, and C. E. Cowan. 1988. Thermochemical Data Used by the FASTCHEM™ Package. Report EA-5872. Electric Power Research Institute, Palo Alto, CA.

Langmuir, I. 1918. The adsorption of gases on plane surfaces of glass, mica, and platinum. *J. Am. Chem. Soc.* 40:1361–1403.

Lasaga, A. C. 1984. Chemical kinetics of water-rock interactions. *J. Geophys. Res.* 89:4009–4025.

Lewis, F. M., C. I. Voss, and J. Rubin. 1986. Numerical simulation of advective-dispersive multisolute transport with sorption, ion exchange and equilibrium chemistry. U.S. Geological Survey Water Resources Investigation. Report No. 86-4022.

Lewis, F. M., C. I. Voss, and J. Rubin. 1987. Solute transport with equilibrium aqueous complexation and either sorption or ion exchange: simulation methodology and applications. *J. Hydrol.* 90:81–115.

Lichtner, P. C. 1985. Continuum model for simultaneous chemical reactions and mass transport in hydrothermal systems. *Geochim. Cosmochim. Acta* 49:779–800.

Lichtner, P. C. 1988. The quasi-stationary state approximation to coupled mass transport and fluid-rock interaction in a porous medium. *Geochim. Cosmochim. Acta* 52:143–165.

Liu, C. W. and T. N. Narasimhan. 1989a. Redox-controlled multiple-species reactive chemical transport. 1. Model development. *Water Resour. Res.* 25(5):869–882.

Liu, C. W. and T. N. Narasimhan. 1989b. Redox-controlled multiple-species reactive chemical transport. 2. Verification and application. *Water Resour. Res.* 25(5):883–910.

Mangold, D. C. and C.-F. Tsang. 1991. A summary of subsurface hydrological and hydrochemical models. *Rev. Geophys.* 29(1):51–79.

Marle, C. M. 1982. On macroscopic equations governing multiphase flow with diffusion and chemical reactions in porous media. *Int. J. Eng. Sci.* 20:643–662.

Marquardt, D. W. 1963. An algorithm for least-squares estimation of nonlinear parameters. *J. Soc. Ind. Appl. Math.* 11:431–441.

Martell, A. E. and R. M. Smith. 1976–1977. Critical Stability Constants. Plenum Press, New York.

Miller, C. T. 1987. Groundwater quality. *J. Water Pollut. Control Fed.* 59(6):513–531.

Miller, C. W. and L. V. Benson. 1983. Simulation of solute transport in a chemically reactive heterogeneous system. *Water Resour. Res.* 19(2):381–391.

Millero, F. J. and D. R. Schreiber. 1982. Use of the ion pairing model to estimate activity coefficients of the ionic components of natural waters. *Am. J. Sci.* 282(9):1508–1540.

Morel, F. and J. Morgan. 1972. A numerical method for computing equilibria in aqueous chemical systems. *Environ. Sci. Tech.* 6:58–67.

Morrey, J. R. 1981. EQUILIB — A Computer Program to Predict Complex Aqueous Equilibria Between 0°C and 300°C. Computers in Chemistry Symposium, National A.C.S. Meeting, Atlanta, GA, March 29.

Morrey, J. R. 1988. FASTCHEM™ Package: User's Guide to ECHEM, the Equilibrium Geochemistry Code, Vol. 4. Report EA-5870-CCM. Electric Power Research Institute, Palo Alto, CA.

Morrey, J. R., C. T. Kincaid, and C. J. Hostetler. 1986. Geohydrochemical Models for Solute Migration: Evaluation of Selected Computer Codes, Vol. 3. Report EA-3417. Electric Power Research Institute, Palo Alto, CA.

Nielsen, O. J., P. Bo, and L. Carlsen. 1985. COLUMN2 —A Computer Program for Simulating Migration. Report No. RISØ-R-514. Risø National Laboratory, Roskilde, Denmark.

Nienhuis, P. and C. A. J. Appelo. 1988. Adaption of PHREEQE for Use in a Mixing-Cell Flowtube: PHREEQM, User Guide. Free University, Amsterdam, The Netherlands.

Nieva, D., R. M. Barragán, M. Cathelineau, E. González, G. Izquierdo, R. Oliver, E. Portugal, E. Santoyo, and M. Verma. 1987. Estudio de la estructura hidrológica y de la distribución de parámetros fisicoquimicos del yacimiento de Los Azufres, Mich.: Segunda Fase, Rep. IIE/11/2011/1 04/F, 159 pp.

Noorishad, J., C. L. Carnahan, and L. V. Benson. 1987. Development of the Non-Equilibrium Reactive Chemical Transport Code CHMTRNS. Report LBL-22361. Earth Sciences Division, Lawrence Berkeley Laboratory, University of California, Berkeley, CA.

Ortoleva, P., E. Merino, C. Moore, and J. Chadam. 1987. Geochemical self-organization. I. Reaction-transport feedbacks and modeling approach. *Am. J. Sci.* 287:979–1007.

Panigatti, J. L. 1970. The Influence of Soil Textural Stratification on Chloride and Fluometuron Movement. Unpublished M.S. thesis, Oklahoma State University, Stillwater, OK.

Parker, D. R., T. B. Kinraide, and L. W. Zelazny. 1988. Aluminum speciation and phytotoxicity in dilute hydroxy-aluminum solutions. *Soil Sci. Soc. Am. J.* 52(2):438–444.

Parker, J. C. 1984. Analysis of solute transport in column tracer studies. *Soil Sci. Soc. Am. J.* 48:719–724.

Parker, J. C. and M. Th. van Genuchten. 1984a. Determining Transport Parameters from Laboratory and Field Tracer Experiments. Bulletin 84-3. Virginia Agricultural Experimental Station, Blacksburg, VA.

Parker, J. C. and M. Th. van Genuchten. 1984b. Flux-averaged and volume-averaged concentration in continuum approaches to solute transport. *Water Resour. Res.* 20(7):866–872.

Parker, J. C. and M. Th. van Genuchten. 1986. Reply. *Water Resour. Res.* 22(7):1159–1160.

Parkhurst, D. L., D. C. Thorstenson, and L. M. Plummer. 1980. U.S. Geological Survey Water Resources Investigation 80-96, NTIS Accession No. PB8I 167801. November.

Persaud, N., J. V. Giraldez, and A. C. Chang. 1985. Monte-Carlo simulation of non-interacting solute transport in a spatially heterogeneous soil. *Soil Sci. Soc. Am. J.* 49:562–568.

Peterson, S. R. and K. M. Krupka. 1981. Contact of Clay Liner Materials with Acidic Tailings Solution. II. Geochemical Modeling. Proc. of the Fourth Symposium on Uranium Mill Tailings Management, Colorado State University, Fort Collins, CO, pp. 609–626.

Pitzer, K. S. 1973. Thermodynamics of electrolytes. I. Theoretical basis and general equations. *J. Phys. Chem.* 77(2):268–277.

Pitzer, K. S. and G. Mayorga. 1973. Thermodynamic of electrolytes. II. Activity and osmotic coefficients for strong electrolytes with one or both ions univalent. *J. Phys. Chem.* 77(19):2300–2308.

Plummer, L. N., B. F. Jones, and A. H. Truesdell. 1976. WATEQF — A FORTRAN IV version of WATEQ, a computer program for calculating chemical equilibria of natural waters. *Water Resour. Invest. U.S. Geol. Surv.* 76.

Rao, P. S. C., J. M. Davidson, R. E. Jessup, and H. M. Selim. 1979. Evaluation of conceptual models for describing non-equilibrium adsorption-desorption of pesticide during steady flow in soils. *Soil Sci. Soc. Am. J.* 43:22–28.

Ratkowsky, D. A. 1986. A statistical study of seven curves for describing the sorption of phosphate by soil. *J. Soil Sci.* 37(2):183–189.

Reed, M. H. 1982. Calculation of multicomponent chemical equilibria and reaction processes in systems involving minerals, gases and an aqueous phase. *Geochim. Cosmochim. Acta* 46:513–528.

Reed, M. H. 1992. Computer Modeling of Chemical Processes in Geothermal Systems: Examples of Boiling, Mixing and Water-Rock Reaction. Applications of Geochemistry in Geothermal Reservoir Development. United Nations Development Program. United Nations Institute for Training and Research (in press).

Reed, M. H. and N. F. Spycher. 1989. SOLTHERM Data Base. A Compilation of Thermodynamic Data from 25 to 300°C for Aqueous Species, Minerals and Gases. Geological Science Department, University of Oregon, Eugene, OR.

Reeves, H. and D. J. Kirkner, 1988. Multicomponent mass transport with homogeneous and herterogeneous chemical reactions: effect of chemistry on the choice of numerical algorithm. 2. Numerical results. *Water Resour. Res.* 24(10):1730–1739.

Richards, L. A. 1931. Capillary conduction of liquids through porous mediums. *Physics* 1:318–333.

Roberts, P. V., M. N. Goltz, and D. M. Mackay. 1986. A natural gradient experiment on solute transport in a sand aquifer. 3. Retardation estimates and mass balances for organic solutes. *Water Resour. Res.* 22(13):2047–2058.

Robie, R. A., B. S. Hemingway, and J. R. Fisher. 1978. Thermodynamic properties of minerals and elated substances at 298.15K and 1 bar (10^5 pascals) pressure and at higher temperatures. *U.S. Geol. Surv. Bull.* 1452.

Rowe, R. K. and J. R. Booker. 1985a. 1-D pollutant migration in soils of finite depth. *J. Geotech. Eng. ASCE* 111(4):479–499.

Rowe, R. K. and J. R. Booker. 1985b. Two-dimensional pollutant migration in soils of finite depth. *Can. Geotech. J.* 22(4):429–436.

Rowe, R. K., J. R. Booker, and C. J. Caers. 1984. POLLUTE 3.3: 1-D Pollutant Migration through a Non-Homogeneous Soil. User Manual. Report 84-13. The Systems Analysis, Control and Design Activity, Faculty of Engineering Science, The University of Western Ontario, London, Ontario, Canada.

Sadiq, M. and W. L. Lindsey. 1979. Selection of standard free energies of formation for use in soil chemistry. Technical Bulletin 134. Colorado State University, Fort Collins, CO.

Schay, G. 1969. Adsorption of solutions of nonelectrolytes. In: Surface and Colloid Science. E. Matijević (Ed.). John Wiley & Sons, New York, pp. 155–211.

Schulz, H. D. and E. J. Reardon. 1983. A combined mixing cell/analytical model to describe two-dimensional reactive solute transport for unidirectional groundwater flow. *Water Resour. Res.* 19(2):493–502.

Selim, H. M., J. M. Davidson, and R. S. Mansell. 1976. Evaluation of a two-site adsorption-desorption model for describing solute transport in soils. Proc. of the 1976 Summer Computer Simulation Conference, Washington, D.C., July, pp. 444–448.

Shamir, U. Y. and D. R. F. Harleman. 1967. Dispersion in layered porous media. *J. Hydraul. Div. Am. Soc. Civ. Eng.* 93:236–260.

Shapiro, R. E. and M. Fried. 1959. Relative release and retentiveness of soil phosphates. *Soil Sci. Soc. Am. Proc.* 23:195–198.

Sibbesen, E. 1981. Some new equations to describe phosphate sorption by soils. *J. Soil Sci.* 32(1):67–74.

Sillen, L. G. 1967. Master variables and activity scales. *Adv. Chem. Ser.* 67:45–56.

Simmons, C. S., C. T. Kincaid, and A. E. Reisenauer. 1986. A Simplified Model for Radioactive Contaminant Transport: The TRANSS Code. Report No. PNL-6029. Pacific Northwest Laboratory, Richland, WA.

Sposito, G. 1979. Derivation of the Langmuir equation for ion exchange reactions in soils. *Soil Sci. Soc. Am. J.* 43(1):197–198.

Sposito, G. 1980. Derivation of the Freundlich equation for ion exchange reactions in soils. *Soil Sci. Soc. Am. J.* 44:652–654.

Sposito, G. 1981. The Thermodynamics of Soil Solutions. Oxford University Press, New York.

Sposito, G. 1982. On the use of the Langmuir equation in the interpretation of "adsorption" phenomena. II. The "two-surface" Langmuir equation. *Soil Sci. Soc. Am. J.* 46(6):1147–1152.

Sposito, G. 1984a. Chemical models of inorganic pollutants in soils. *CRC Crit. Rev. Environ. Control* 15(1):1–24.

Sposito, G. 1984b. The future of an illusion: ion activities in solution. *Soil Sci. Soc. Am. J.* 48:531–536.

Sposito, G. 1984c. The Surface Chemistry of Soils. Oxford University Press, New York.

Sposito, G. 1989. The Chemistry of Soils. Oxford University Press, New York.

Sposito, G. and D. A. Barry. 1987. On the Dagan model of solute transport in groundwater: foundational aspects. *Water Resour. Res.* 23(10):1867–1875.

Sposito, G. and J. Coves. 1988. SOILCHEM: A Computer Program for the Calculation of Chemical Speciation in Soils. Report. Kearney Foundation, University of California, Riverside and Berkeley, CA.

Sposito, G. and W. A. Jury. 1988. The lifetime probability density function for solute movement in the subsurface zone. *J. Hydrol.* 102(1–4):503–518.

Sposito, G. and S. V. Mattigod. 1980. GEOCHEM: A Computer Program for the Calculation of Chemical Equilibria in Soil Solutions and Other Natural Water Systems. Kearney Foundation, University of California, Riverside and Berkeley, CA.

Summers, K. V., S. A. Gherini, M. M. Lang, M. J. Ungs, and K. J. Wilkinson. 1989. MYGRT Code Version 2.0: An IBM Code for Simulating Migration of Organic and Inorganic Chemicals in Groundwater. Report EN-6531. Electric Power Research Institute, Palo Alto, CA.

Talbot, A. 1979. The accurate numerical integration of Laplace transforms. *J. Inst. Math. Appl.* 23:97–120.

Tan, K. H. 1982. Principles of Soil Chemistry. Marcel Dekker, New York.

Theis, T. L., D. J. Kirkner, and A. A. Jennings. 1982. Multi-Solute Subsurface Modeling for Energy Solid Wastes. COO-10253-3. Department of Civil Engineering, University of Notre Dame, Notre Dame, IN.

Torgensen, T. 1991. Crustal fluid flow: Continuous or episodic? *Eos Trans. Am. Geophys. Union* 72(3):18–19.

Truesdell, A. H. and B. F. Jones. 1974. WATEQ, a computer program for calculating chemical equilibria of natural waters. *J. Res. U.S. Geol. Surv.* 2(2):233–248.

Valocchi, A. J., R. L. Street, and P. V. Roberts. 1983. Transport of ion-exchanging solutes in ground water: chromatographic theory and field simulation. *Water Resour. Res.* 17(5):1517–1527.

van der Zee, S. E. A. T. M. 1990. Analysis of solute redistribution in a heterogeneous field. *Water Resour. Res.* 26:273–278.

van der Zee, S. E. A. T. M. and W. H. van Riemsdijk. 1986. Transport of phosphate in a heterogeneous field. *Transp. Porous Media* 1:339–359.

van Gaans, P. F. M. 1989. WATEQX — a restructured, generalized, and extended FORTRAN 77 computer code and database format for the WATEQ aqueous chemical model for element speciation and mineral saturation, for use on personal computers or mainframes. *Comput. Geosci.* 15(6):843–887.

van Genuchten, M. Th. 1974. Mass transfer studies in sorbing porous media. Ph.D. thesis. New Mexico State University, Las Cruces, NM.

van Genuchten, M. Th. 1979. A closed-form equation for predicting the hydraulic conductivity of unsaturated soils. *Soil Sci. Soc. Am. J.* 44:892–898.

van Genuchten, M. Th. 1981. Non-equilibrium transport parameters from miscible displacement experiments. Research Report No. 119. U.S. Salinity Laboratory, Riverside, CA.

van Genuchten, M. Th. and R. W. Cleary. 1982. Movement of solutes in soil: computer-simulated and laboratory results. In: Soil Chemistry. B. Physico-Chemical Models. G. Bolt (Ed.). Elsevier, Amsterdam, The Netherlands, pp. 349–386.

Voss, C. I. 1984. SUTRA: a finite-element simulation model for saturated-unsaturated, fluid-density-dependent ground-water flow with energy transport or chemically-reactive single species solute transport. *U.S. Geol. Surv. Water Resour. Invest. Rep.* 84–4369.

Walsh, M. P., S. L. Bryant, R. S. Schechter, and L. W. Lake. 1984. Precipitation and dissolution of solids attending flow through porous media. *AIChE J.* 30:317–328.

Wigley, T. M. L. 1977. WATSPEC: A Computer Program for Determining the Equilibrium Speciation of Aqueous Solutions. Tech. Bull. 20, British Geomorph. Res. Group. Exeter, England.

Wolery, T. J. 1979. Calculation of Chemical Equilibrium between Aqueous Solution and Minerals: The EQ3/6 Software Package. UCR-52658. Lawrence Livermore Laboratory, Livermore, CA.

Wolery, T. J. 1983. EQ3NR: A Computer Program for Geochemical Aqueous Speciation-Solubility Calculations, User's Guide and Documentation. Report UCRL-53414. Lawrence Livermore Laboratory, Livermore, CA.

Young, D. M. and A. D. Crowell. 1962. Physical Adsorption of Gases. Butterworths, London.

CHAPTER 4

Modeling Salt Transport in the Landscape

Peter Thorburn, Roger Shaw, and Ian Gordon

4.1 INTRODUCTION

Movement of solutes through the vadose zone, to groundwater, is of increasing concern as greater pressure is put on our environment. Solutes of particular concern are pesticides or heavy metals, which are substantially sorbed onto soil particles. However, there are other chemicals that may contaminate groundwater resources. Large quantities of nitrogen are applied in many agricultural areas, while many irrigation waters contain high solute concentrations, often chlorides. Most commonly nitrate and chloride are not significantly sorbed by soils, so they move freely (once below the root zone in the case of nitrate) downward with soil water.

The effects of salt movement in soils were well recognized in ancient Mesopotamia where the capillary rise of water from shallow water tables carried dissolved salts to the soil surface with significant effects on agriculture (Jacobsen and Adams, 1958). Warington (1900) was one of the early writers this century to describe salt movement in soils. He described salt movement as having both a diffusive component and a mass flow component with water movement. He was also aware of general piston flow through the matrix and differential flow of water through macropores which he related to differing salt concentrations in the drainage waters.

The movement of dissolved salts in rivers and streams is also an important consideration in surface water-quality decisions. The growing need for information on river salt loads has been spurred on by increasing human impact upon river systems and the necessity of maintaining water quality. Interest in the transport of dissolved

ISBN 0-87371-747-3

salts by rivers can be traced back for over 100 years (Walling and Webb, 1983). In most cases early work focused on chemical analyses of individual water samples rather than attempting to document the salt loads transported during specific periods of time. This is still largely true of more recent studies, because more continuous monitoring of salinity is still not nearly as common as the estimation of discharge.

This chapter, which actually consists of three subchapters, focuses on the types of models that have proved practically useful in managing salt in landscapes. In the near surface, where salt concentration distribution is of particular significance to vegetation, most use has been made of one-dimensional models. Section 4.2 provides the theoretical framework for describing and predicting salt movement in the vadose zone above the water table. Especially at the scale of rooting depths there is interest in the spatial and temporal distribution of salt concentration.

At the scale of landscape elements, such as small valleys, change may not be so rapid. The onset of drought, or a series of wet years, may lead to significant change in the salt balance. However, the longer-term steady-state salt balance for the landscape element can still provide a most useful reference condition from which shorter-term departures can be understood.

If the changes in the evaporation regime of a valley are substantial, for example due to the clearing of timber, then this may lead to departure from a prior steady state toward another steady state with a rather different salt balance. This kind of change is commonly associated with salinity problems. Section 4.3 illustrates how salt balance models that refer to steady-state situations can be of great assistance in land management issues where salinity is a constraint of concern.

Section 4.4 shows how an understanding of salt balance in streams can be of use in aiding catchment management decisions. Here the concept of a steady state as examined in detail in Section 4.3 is applied to a stream or river draining a catchment. Such analysis can be helpful in indicating the sensitivity of the salt load of a stream draining a catchment to hydrologic perturbation, such as that that results from an extensive change in vegetation type in a catchment.

Many of the references in this chapter cite Australian experience. Australia has the highest proportion of salt-affected land of any continent (Szabolcs, 1986). For this reason considerable research work has been carried out on various aspects of salt movement in the soil profile, and in landscape elements and catchments. Such studies have been mainly carried out in the salt-affected regions of Queensland with the direct involvement of the authors.

This chapter consists of three main subsections, each prepared by one of the three contributors. Section 4.2 on the one-dimensional non-steady-state solute movement in the vadose zone has been written by Peter Thorburn. Section 4.3 on the one-dimensional steady-state salt transport and Section 4.4 on the use of stream salt mass balance in catchment management have been prepared by Roger Shaw and Ian Gordon, respectively.

4.2 ONE-DIMENSIONAL NON-STEADY-STATE SOLUTE MOVEMENT IN THE VADOSE ZONE

4.2.1 Introduction

While steady-state models have a role to play in environmental management, they cannot provide predictions of solute movement under conditions of variable land management, or where solutes are stored in or leached from the soil. Transient models are required under these conditions. However, transient models may be complex, requiring data that may be difficult to obtain by users in many circumstances. Conversely, simpler transient mass balance models will be more readily applicable, but will give less detailed predictions. Thus it is necessary to be aware of the range of transient models available, and choose that which represents the best compromise between data limitations and detail of prediction required.

Many options are available to model solute movement in the vadose zone. These options are numerical or analytical solutions of the convection-dispersion equation, stochastic models of solute movement, theories that are discontinuous with depth (i.e., based on layer approaches), or transient theories based on mass balance of solutes within the root zone that do not describe the depth distribution of solute in soils. Often the different models have been developed for specific applications, and so reflect the special requirements of that application. Also, other factors such as soil structure or crop uptake can affect solute movement and different models have been developed to account for these factors. In this section, the different basic transient approaches to modeling movement of nonsorbed solutes in soils are identified and briefly described, and the relevance of each basic approach for widespread application in field soils discussed. Emphasis is given to models that can be readily applied under a wide range of field conditions. Application of these models will be mainly considered where chloride and nitrate movement is of interest at the base of the root zone, as it is under these conditions that most work has been reported, and it is in the root zone where greatest variations in the concentrations of solutes occur.

4.2.2 Models Based on the Convection-Dispersion Equation

The processes involved in solute movement are commonly described by the convection-dispersion equation (CDE):

$$\partial c / \partial t = D\, \partial^2 c / \partial z^2 - v\, \partial c / \partial z \tag{1}$$

where c is solute concentration, t is time, D is the dispersion coefficient, and v is the pore water velocity, given as the water flux (q) divided by the volumetric water content (θ).

The coefficient D accounts for both hydrodynamic dispersion and molecular diffusion, and its value is given by (De Smedt and Wierenga, 1978b)

$$D = D_0 + \varepsilon v \tag{2}$$

where ε is the coefficient of solute dispersivity in soil, and D_0 is the molecular diffusion coefficient.

Solutions to the CDE are obtained numerically or, under certain conditions, analytically.

4.2.2.1 Numerical Solutions of the Convection-Dispersion Equation

This approach has been successfully used to describe solute movement in soils (Bresler and Hanks, 1969; Bresler, 1973; Kirda et al., 1973, 1974; Tillotson and Wagenet, 1982) and has been modified to include other processes such as chemical adsorption (e.g., Robbins et al., 1980; Wagenet and Hutson, 1984).

This method of describing soil solute movement suffers limitations because of spatial variability of model parameters. For example, hydraulic conductivity, K (Rogowski, 1972; Nielsen et al., 1973), D, and v (Biggar and Nielsen, 1976; van de Pol et al., 1977, Russo and Bresler, 1981) are often variable and log-normally distributed in "uniform" soils. Thus their characterization requires major field experimentation. Also, relationships between K, θ, and the soil matric potential (ψ) are assumed to be unique, which is usually not the case, due to hysteresis. Thus, although numerical solutions of the CDE are theoretically sound, unless their inputs are properly characterized their predictions may not be accurate (Addiscott and Wagenet, 1985).

4.2.2.2 Analytical Solutions of the Convection-Dispersion Equation

Analytical solutions for Equation (1) have been obtained under many boundary conditions (van Genuchten and Alves, 1982). They have been found to adequately describe solute transport in many laboratory studies (e.g., Nielsen and Biggar, 1962; Rose and Passioura, 1971; Gupta and Greenkorn, 1974; Cameron and Klute, 1977; Yule and Gardner, 1978), and are generally simpler than numerical solutions, in terms of both input data required and ease of computing (Wierenga, 1977; Addiscott and Wagenet, 1985). Analytical solutions of the CDE however, generally assume q and θ to be constant with both time and depth. These conditions are rarely met in field soils. For example, rainfall or irrigation can be of variable intensity and infiltration rates are higher in dry soils than in wet. Also, θ can vary with soil depth, during initial wetting of soils, and in soils with textural changes within the profile. Thus the utility of analytical solutions of the CDE was thought to be severely restricted in field soils, and the use of numerical solutions became popular.

However, Cassel et al. (1975) noted that when depth-varying values of θ were replaced by a depth-averaged value in a numerical solution of the CDE, agreement with experimental data was good considering variations in solute movement between replicate cores. Subsequent experiments have investigated the effect of assuming average or steady values of θ (Wierenga, 1977; Selim et al., 1977; De Smedt and Wierenga, 1978a), q (Wierenga, 1977; De Smedt and Wierenga, 1978b), and D (De Smedt and Wierenga, 1978a) on solute transport in situations where these parameters varied with depth and/or time. These experiments showed that the accuracy of predictions of solute transport was little affected by the steady-state assumptions. Thus, average values of θ, q, and D can replace nonsteady values in calculating solute movement. Although these results were obtained in laboratory studies, these findings are likely to be general principles that should be applicable in field soils as well (Rose et al., 1982b).

The application of these findings to field soils allows analytical solutions of the CDE to be used to calculate solute movement under transient conditions, these solutions being simpler to use than numerical solutions (as stated above). Additionally, only θ, and not detailed soil hydraulic properties, needs to be determined to characterize solute movement. Generally θ is much less spatially variable than the hydraulic functions $K(\theta)$ and $\psi(\theta)$, and tends to be normally distributed in field soils (McIntyre and Tanner, 1959; Rogowski, 1972; Nielsen et al., 1973; Biggar and Nielsen, 1976). θ is also easily determined in contrast to the hydraulic functions (Wierenga, 1977; Rose et al., 1982c).

Analytical solutions of the CDE can be further simplified if the movement of a solute concentration peak in soil is known. The movement of the peak provides a *Lagrangian* coordinate system about which only dispersion needs to be predicted from the CDE (Gardner, 1965; Rose et al., 1982b). Theories for describing the movement of a solute peak in soil and solutions to Equation (1) for solute dispersion about a known peak have been developed by De Smedt and Wierenga (1978b) and Rose et al. (1982a–c), and have been successfully applied in field situations. Because of their simplicity, these models based on analytical solutions to the CDE have potential as management tools for use in field soils. They will be considered in more detail in the following section.

4.2.2.3 Prediction of Peak Movement and Dispersion

De Smedt and Wierenga (1978b) developed a model for movement of a concentration peak of a nonadsorbed, nonreactive solute through soils for constant q. Subsequently, Rose et al. (1982a,c) relieved this requirement for constant q, using a mass balance approach. The movement of the peak of a solute concentration profile through a uniform soil is described by a very simple but moderately accurate solution to Equation (1), assuming that there is no preferential movement of water through soils or wetting front instability effects, the soil drains to a constant water content following infiltration, such as "field capacity" (θ_{fc}), no significant soil-solute interactions or chemical transformations occur, crop water uptake is uniform with depth to the bottom of the root zone, and the profile is at field capacity below the root zone.

Rose et al. (1982a,c) defined the depth (α) beneath the soil surface of the peak concentration of solute as the depth at which $dc/dz = 0$. Although strictly α is that depth where $\partial c/\partial z = 0$, that is, including the effects of dispersion, Rose et al. (1982c) illustrated that the neglect of dispersion in defining α would only lead to an overestimate of α of ~ 10%. They considered this error acceptable within the context of the application of the theory.

Movement of α within a profile is calculated assuming that the solute moves through the soil displaced by the wetting front of successive water applications (irrigation and rainfall), and that evapotranspiration occurs between infiltration events. The depth of α from the soil surface (downward positive) in each event is calculated in finite time elements. For the nth time element,

$$\alpha_n = \alpha_{n-1} + \Delta W_{n,z=\alpha_{n-1}} / \theta_{\text{fc}} \tag{3}$$

where $\Delta W_{n,z=\alpha_{n-1}}$ is the equivalent ponded depth of water that passed below the previous position of the solute peak ($z=\alpha_{n-1}$), and θ_{fc} is assumed constant with soil depth. A general expression for $\Delta W_{n,z=\alpha_{n-1}}$ is

$$\Delta W_{n,z=\alpha_{n-1}} = I_n \Delta t - \gamma\left(\theta_{\text{fc}} - \theta_{n-1} + \kappa\, E_n \Delta t / R_n\right) \tag{4}$$

where I is the infiltration rate, Δt is the length of the time element, γ is the depth over which the soil water deficit must be replenished for peak movement to occur, θ_{n-1} is the volumetric moisture content at the end of the previous time increment, κ is the proportion of evapotranspiration occurring before infiltration, E_n is the evapotranspiration rate, and R_n is the depth of rooting. $I\Delta t$ and $E\Delta t$ are expressed as equivalent ponded depths of water.

If peak movement occurs, two extreme boundary conditions are possible: first, that infiltration is complete before evapotranspiration, and second the reverse. These conditions are taken into account in Equation (4) by selecting the value of κ. If infiltration is assumed complete before evapotranspiration, then $\kappa = 0$. For the reverse condition, $\kappa = 1$. If the value of κ cannot be chosen with confidence (for example, from irrigation management data), calculations can be made using the extreme values (0 and 1) to encompass the uncertainty in the prediction of peak movement. The value of γ depends on the relationship between α_{n-1} and R_n. If $\alpha_{n-1} < R_n$, $\gamma = \alpha_{n-1}$, but if $\alpha_{n-1} > R_n$, $\gamma = R_n$. The previous profile water content, θ_{n-1}, is calculated from water balance principles (Rose et al., 1982a).

Thus, the data required for calculation of the displacement of a solute concentration peak are a knowledge of irrigation application events, rainfall, evapotranspiration, and variation in crop rooting depth with time.

Based on earlier solutions of Warrick et al. (1971, 1972), De Smedt and Wierenga (1978b) solved Equation (1) for constant flux boundary conditions and a step *increase* in input concentration (i.e., no solute/solute) of nonadsorbed, nonreactive solute. De Smedt and Wierenga (1978b) compared predictions of chloride movement from this solution to Equation (1) with predictions from a numerical solution of the CDE, and found close agreement. Recently, Elrick et al. (1987) showed how a series of successively simpler analytical solutions to the CDE (the "simplest" approximation being that of De Smedt and Wierenga, 1978b) can be derived from an exact solution (of Smiles et al., 1978) by making assumptions about soil water movement. They examined the different chloride profiles derived from the range of simplified solutions, and found that all were good approximations of the exact solution, further supporting use of the approach in field soils.

Rose et al. (1982b) defined the depth of penetration of the step *decrease* in input nonadsorbed, nonreactive solute (i.e., solute/no solute), β, as

$$\beta = \alpha - \tau \tag{5}$$

where τ is the depth of the solute pulse.

They then used β to solve Equation (1) for constant flux boundary conditions and a step decrease in solute concentration, and combined this solution with that of De Smedt and Wierenga (1978b), employing the principle of superposition (Rose, 1977) to give the resultant solute concentration profile:

$$c/c_0 = 1/2\left\{\operatorname{erfc}\left[(z-\alpha)/2(D_0t+\alpha\varepsilon)^{\frac{1}{2}}\right] - \operatorname{erfc}\left[(z-\beta)/2(D_0t+\beta\varepsilon)^{\frac{1}{2}}\right]\right\} \tag{6}$$

where c_0 is the initial solute concentration.

In their application of Equation (6), Barry et al. (1985) approximated v in Equation (1) by the mean pore water velocity, $v_m = \Delta\alpha/\Delta t$. Values for τ, ε, and D_0 are required to calculate solute concentrations. Where solute is applied as irrigation water (or as readily soluble fertilizer), Barry et al. (1985) showed that

$$\tau = I_n \Delta t / \theta_{fc} \tag{7}$$

The value of ε is not known *a priori*, so must be determined, usually from fitting predicted to measured data. Rowell et al. (1967) found D_0 to be ~ 60 mm^2 day^{-1} for chloride in a wet sandy clay loam, although D_0 is linearly related to θ (Nye and Tinker, 1977). Similar values of D_0 have been used in previously reported applications of Equation (6).

Where crop solute uptake is of interest, for instance when considering nitrate movement, detailed theory is available for its description (Nye and Tinker, 1977). However, a simple field-applicable approximation method for accounting for solute uptake was presented by Rose et al. (1982b). If the amount of solute taken up by the crop (u) is known as a function of time, a factor (Φ) to account for solute uptake can be calculated from

$$\Phi = (c_n - u)/c_n \tag{8}$$

where c_n is the mass of solute (e.g., inorganic nitrogen) in the soil profile.

The right-hand side of Equation (6) is multiplied by Φ to calculate the soil solute concentration. This method of accounting for crop solute uptake was found to adequately describe field nitrate data by Rose et al. (1982b) and Barry et al. (1985).

Smith et al. (1984) developed a more detailed modification of Equation (6) to account for crop solute uptake. They based their modification on a linear relationship between crop uptake and increasing solute concentration. They successfully modeled the subsequent distribution of solute applied to the surface of a range of soils. The accuracy of their predictions, however, was similar to those of Rose et al. (1982b) and Barry et al. (1985). Thus there appears to be little gain in adopting this more detailed approach when describing field data.

4.2.2.4 Application of Models Based on the Convection-Dispersion Equation

Although both analytical and numerical solutions of the CDE have successfully described solute movement in many studies, there are some reservations to routine applications of CDE-based models in field soils. As discussed above, values of D and v are variable even in a "uniform" field, and are log-normally distributed. Intensive sampling would thus be required to characterize mean field values of D and v, yet these mean values would be of little practical value for predicting solute movement in the extreme cases (i.e., with large "real" values of D and v). However, these reservations regarding variation in values of D and v are less of a limitation to the field application of simple analytical solutions of the CDE, as they rely on capacity variables. Because of their ease of use and the relatively few limitations for field application, simple analytical solutions of the CDE have been successfully applied in field studies. The results of these studies will be examined.

Field applications of simple analytical solutions of the CDE have centered on Equation (6) (referred to as the PEAK model). Details of these studies are summarized in Table 4.1. In all studies predicted and measured data generally were in good agreement (especially considering the normal variation expected with field measurements), except when applied to flood-irrigated cracking clay soils (Thorburn et al., 1990a).

Rose et al. (1982a–c) and Barry et al. (1985) predicted solute movement below a fixed soil depth, while Cameron and Wild (1982) and Thorburn et al. (1990a) predicted solute profiles arising from a surface application of solute. Nitrate fertilizer was applied a number of times at the two sites modeled by Rose et al. (1982a,b) and Barry et al. (1985). The concentration at depth, the base of a lysimeter, was calculated by adding the effects of each solute application. Barry et al. (1985) found that predictions from the PEAK model were more accurate than those of the numerical model, which accounted for spatial variation in infiltration, used by Jury et al. (1976) with data from one of the sites. Cameron and Wild (1982) applied chloride at one time, followed by four irrigations and/or rainfall events. Solute profiles were more accurately predicted by the PEAK model than two multicompartmental (i.e., "layer") models. Thorburn et al. (1990a) applied the model to eight sites where the solute (chloride) was contained in irrigation water. Each irrigation application resulted in the addition of solute, and the modeled profile was "constructed" by superimposing the profiles arising from each previous irrigation.

The PEAK model did not accurately predict a minor solute peak after the primary peak in the data examined by Rose et al. (1982a,b), or prior to the primary peak in the study of Barry et al. (1985). These errors were attributed to storage and subsequent release of solute within fissured shale bedrock within the lysimeter in the former instance (Rose et al., 1983), and to preferential water flow within the lysimeter in the latter case (Barry et al., 1985). Predicted profiles were substantially lower than measured profiles in three of the sites examined by Thorburn et al. (1990a). The error was attributed to preferential infiltration of low-salinity irrigation water through cracks in the clay soil, which did not leach solute resident in the soil matrix. The preferential infiltration of heavy rainfall also had to be accounted for at two other sites with clay soils (Figure 4.1) examined by Thorburn et al. (1990a).

The range of values of κ, ε, and D_0 used in applications of the PEAK model are listed in Table 4.1. Sensitivity of the model to these variables has been explored in field studies by Cameron and Wild (1982), Barry et al. (1985), and Thorburn et al.

Table 4.1 Details of Reported Applications of the PEAK Model of Rose et al. (1982a,b)

Study	Rose et al. (1982a,b)	Cameron and Wild (1982)	Barry et al. (1985)	Thorburn et al. (1990a)	De Smedt and Wierenga (1978b)[a]
Number of sites	2	1	2	8	
Soil texture	Silty loam and loamy sand	Silty clay and loam	Silty loam and loamy sand	Loam to heavy clay	Silty clay loam
Solute	NO_3^-	Cl^-	NO_3^- and Cl^-	Cl^-	Cl^-
Time increment	Month	Variable	Day	Month	NA[b]
κ (1 or 0)	0	NG[c]	1	1	NA
ε (mm)	120	30 (10–100)[d]	100	10–120[e]	5 (5–20)
D_0 (mm^2 day^{-1})	100	10 (10–100)	100	100 (50–150)	300 (0–300)

[a] Details of the laboratory simulation of De Smedt and Wierenga (1978b) are included for comparison.
[b] NA, Not applicable.
[c] NG, Details not given.
[d] Range of values found to give "acceptable" predictions.
[e] Range across sites.

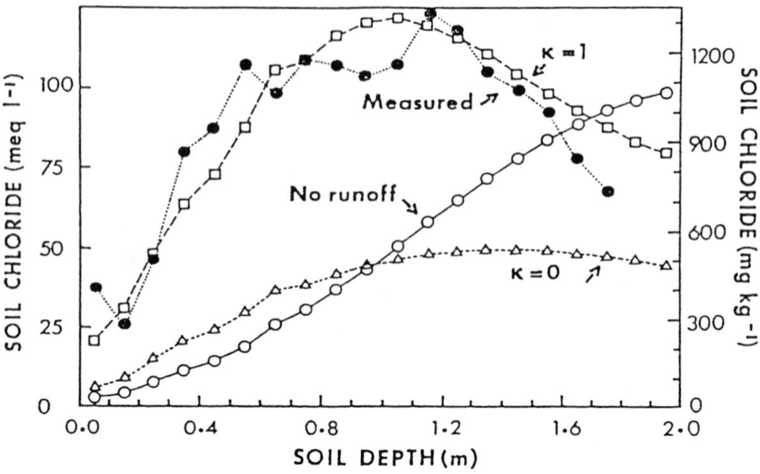

FIGURE 4.1. Soil chloride profiles measured and simulated by the PEAK model assuming evapotranspiration before infiltration ($\kappa = 1$), evapotranspiration after infiltration ($\kappa = 0$), and that all water infiltrates the soil matrix (i.e., no runoff) with $\kappa = 1$. From site DWD-2 of Thorburn et al. (1990a).

(1990a). The model is more responsive to ε than D_0 (Figure 4.2 and Table 4.1), as expected, because D_0 can be considered small (possibly negligible) compared with ε (Passioura, 1971).

In field studies, values of ε have been found one or two orders of magnitude greater than those calculated theoretically (Passioura, 1971). ε may be affected by soil particle or aggregate size and/or soil structure (Passioura, 1971; Biggar and Nielsen, 1976; Cameron and Wild, 1982; Jensen, 1983). Hydrodynamic dispersion theory makes assumptions regarding the pore size distribution in soil (e.g., that there are only two pore sizes, macro and micro; Passioura, 1971) and was tested on porous materials for which these assumptions were reasonably valid (e.g., aggregates of "ignited" clay or crushed porcelain of reasonably uniform size; Passioura and Rose, 1971). In contrast, field soils will have a much greater range and variation in pore sizes, particle diameters, and structure that will result in the larger observed values of ε (Biggar and Nielsen, 1976; Cameron and Wild, 1982).

Development of a means of independently estimating ε would be desirable for application of the PEAK model (Barry et al., 1985; Addiscott and Wagenet, 1985). Values of ε have been related to pore size distribution and aggregate size, but should be independent of pore water velocity (Jensen, 1983). Thorburn et al. (1990a) found that values of ε were significantly correlated (r = 0.90) with the logarithm of the leaching fraction [LF, leaching flux/infiltration (%)] and the logarithm of the leaching flux (r = 0.87) at five of their sites. The regression of leaching fraction and ε (mm) was

$$\varepsilon = 46.6\ \ell n(\mathrm{LF}) - 48.5 \quad (n = 5) \tag{9}$$

Shaw and Thorburn (1985a) found that the leaching fraction was well correlated with soil properties (exchangeable sodium, clay and cation-exchange capacity). They

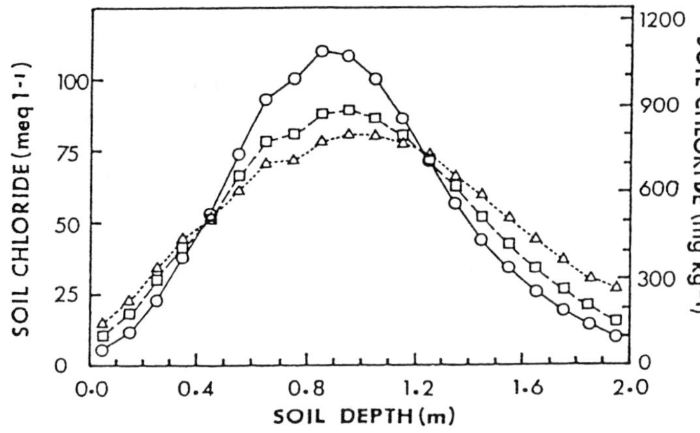

FIGURE 4.2. Soil chloride profiles simulated by the PEAK model using different values of the dispersivity coefficient (ε): ε = 10 mm (○), ε = 70 mm (□), and ε = 120 mm (△). From site DWD-1 of Thorburn et al. (1990a).

hypothesized that those soil properties influenced soil structure and hence soil permeability. The correlation between ε and the leaching fraction found by Thorburn et al. (1990a) thus implies that similar soil structural factors may influence both solute dispersivity and leaching in soils. This is physically plausible. The correlation was based on too few points to be of any real predictive value. However, it indicates an avenue for future investigation.

As values of ε cannot be accurately predicted from soil or irrigation variables, values of ε should be determined by model calibration before application of the PEAK model at a particular site. Where such model calibration is not possible, however, using the values 10 and 120 mm for ε in the simulation of soil chloride profiles appears likely to indicate the range of variation of ε in the root zone of field soils (Table 4.1). Whether ε is more likely to be at the higher or lower extreme of this range at a site may be indicated by the leaching fraction at the site [Equation (9)]. Alternatively, because this apparently extreme range of ε values may give relatively small differences in simulated soil chloride profiles (Figure 4.2), the adoption of a mean value (such as 70 mm) may be adequate for many purposes, even if results are approximate. Certainly variation in ε generally has less effect on predictions than the value chosen for κ, or lack of information on land management or agronomic variables (Barry et al., 1985; Thorburn et al., 1990a).

Adopting extreme values for κ illustrates the range of uncertainty associated with the timing of $I\Delta t$ and $E\Delta t$ (Rose et al., 1982a,c). A value of $\kappa = 0$ was found to provide the most accurate predictions of solute leaching by Rose et al. (1982a,b) for monthly data, while Barry et al. (1985) found a value of $\kappa = 1$ gave the most accurate predictions for a daily time increment. Barry et al. (1985) found with a weekly time increment however, that predictions were similar for both $\kappa = 0$ or 1. Both Barry et al. (1985) and Thorburn et al. (1990a) found a value of $\kappa = 1$ consistent with irrigation management at the sites investigated (e.g., Figure 4.1), and thus suggested its value could often be deduced from knowledge of irrigation practice.

De Smedt and Wierenga (1978b) compared predictions from their "leading edge" solution to Equation (1) against numerical solutions of the CDE for a range of values of ε and D_0. It is difficult to draw conclusions from a comparison of the results of De Smedt and Wierenga's (1978b) predictions and those of field-based studies because of the small times examined (0.1 to 10 d) and the difference in models used by them. It is interesting to note, however, that they used values of ε (and D_0) close to those in the field-based studies, and also found their predictions little affected by similar variations in ε (Table 4.1).

4.2.3 Stochastic Models

Stochastic models generally use either a stochastic description of the spatial variability of parameters input to a mechanistic model (e.g., a CDE-based model), or treat the movement of water and solutes in the vadose zone as a stochastic process. This latter group of models is typified by the transfer function approach.

Dagan and Bresler (1979) developed a model in which solute dispersion was neglected and v was considered to be a random spatial variable, constant in time, whose value was a function of rainfall and irrigation inputs or soil hydraulic properties. They predicted the probabilities of solute reaching a certain concentration at a particular depth and time, the average solute concentration at a depth and time, or the proportion of a field in which the solute concentration was less than a given level at a particular time and depth. Simmons (1982) used a similar approach but allowed for variable v. Further studies by Bresler and Dagan (1979, 1981), Russo and Bresler (1981), and Russo (1984) applied the concept and developed it to include dispersion and the spatial geostatistical relationships between input soil properties. These studies stressed the importance of considering the spatial variability of soil hydraulic properties when applying analytical or numerical solutions to the CDE.

Transfer function models (TFMs) treat the soil as a "black box," and make no assumptions as to the process conveying solute from the soil surface to some depth z. This concept was first applied to soils by Raats (1978a,b), and later Jury (1982) and Jury et al. (1986), who developed the concept to describe movement of solute undergoing physical, chemical, or biological transformations through soil with variable flow in time and space. TFMs have been successfully applied to describe the transport of nonadsorbed, nonreactive solutes through soils in laboratory (core) or field studies by Jury et al. (1982), White et al. (1984), White (1985a,b), Jury and Sposito (1985), White et al. (1986b), and Dyson and White (1987). These experiments have generally found a two-parameter (mean and standard deviation) log-normal distribution of travel times is most useful in describing solute movement.

Solutions to the CDE and TFM have been compared in field soils (Jury, 1984,1988; Jury and Sposito, 1985) and soil cores (Dyson and White, 1987). In field soils, agreement has been good at, and near, the depth of calibration. This finding also has theoretical support (Simmons, 1982). However, at deeper soil depths (e.g., > 3 m) predictions from the two approaches are substantially different, with TFM predictions closer to measured solute concentrations than CDE predictions. In the soil cores studied by Dyson and White (1987), chloride break-through curves were modeled equally well by both the TFM and CDE, provided, however, that the value of v used in the CDE was fitted from experimental data, and not calculated as q/θ_{fc}. This fitting of v was required because of immobile water in the soil cores.

Although stochastic models have been widely used in field applications, they have not generally been applied to an active plant root zone. In a cropped situation, Dudley et al. (1982) and Wagenet and Rao (1983) found lower variability in measured chloride and nitrate concentrations than those predicted by a numerical CDE-based model, given the variability in soil hydraulic properties that had been measured. They concluded that plants affected water fluxes, and hence solute movement, in the root zone, the effect being to "even out" spatial differences in solute and water fluxes. If this is so, models that did not explicitly account for variability in soil hydraulic properties may be more useful for predicting solute movement in cropped soils than otherwise thought. Additionally, these models have not been developed to the simplicity of the PEAK model, possibly restricting their routine application by advisory agencies. TFM models do, however, provide superior predictions of solute movement in the absence of plant roots (i.e., below the root zone).

4.2.4 Layer Models

As the name implies, layer models are discontinuous with soil depth, dividing the soil arbitrarily into a number of finite layers. Many of these models are derived from the application of plate chromatography theory to leaching in soils (van der Molen, 1956; Thomas and Coleman, 1959). The two most widely applied models are those of Burns (1975) and Addiscott (1977), and their subsequent modifications. The Burns and Addiscott models gave good predictions of experimental data in studies of Cameron and Wild (1982), Nicholls et al. (1982), and Khanif et al. (1984). To obtain acceptable predictions with these models, however, Cameron and Wild (1982) had to partition infiltration events arbitrarily into 5-mm increments to satisfy assumptions concerning residence of water within soil layers. The disadvantage of the layer models is their reliance on artificial soil processes, in assuming that the soil is discontinuous with depth. For example, layer models failed to give acceptable predictions in the studies of Barraclough et al. (1983) and Madan Kumar and Oswal (1984) because model assumptions were not met.

4.2.5 Transient Mass Balance Models

Mass balance models do not describe the depth distribution of solute within the root zone. Rather, solute concentration in the whole root zone is predicted as a function of mass inputs and outputs. Pandey and Gupta (1978) developed a model to predict leaching fraction values that will occur when solute is leached by low-salinity irrigation water from a soil with a uniform depth distribution of salinity, or when solute is added to an initially solute-free soil by irrigation. The model is simple, requiring only soil and irrigation water salinity, the volume of water applied to the soil, and soil porosity. However, its assumptions of either solute-free soils, or soils with uniformly depth-distributed solutes, are often not met (e.g., Rose et al., 1979; Thorburn et al., 1990a,b), limiting its application.

In slowly permeable soils long times are required to reach steady-state conditions. Rose et al. (1979) realized that there was a need for a simple theory to describe the change in solute concentrations with time in these soils so the management could be altered to avoid the occurrence of deleteriously high soil solute concentrations. They derived such a model (termed the SODICS model) based on conservation of solute mass within the root zone.

Conservation of solute mass within a soil layer, expressed differentially, gives

$$z\theta_\ell \, d\bar{c}/dt = Ic_i - Ls_z \tag{10}$$

where z is depth, θ_ℓ is the water content at which leaching occurs, $\bar{c}$ is the average solute concentration of the soil water to depth z, c_i is the solute concentration of the infiltrating water, L is the leaching flux past depth z, and $\bar{c}_z$ is the average solute concentration of the soil water at depth z.

A parameter (λ) can be defined that describes the relationship between $\bar{c}$ and the solute concentration of water leaching from the root zone (i.e., c_z):

$$\lambda = c_z/\bar{c} \tag{11}$$

Using this parameter, the solution of Equation (10) is (Rose et al., 1979)

$$\bar{c}_t = \bar{c}_0 + \left(Ic_i/L\lambda - \bar{c}_0\right)\left(1 - \exp\left[-\left(L\lambda / z\theta_\ell\right)t\right]\right) \tag{12}$$

where $\bar{c}_0$ is the value of $\bar{c}$ at time $t = 0$.

With measurements of soil solute concentration profiles at two times, a knowledge of I and c_i between those times, and known values of z and θ_ℓ L becomes the only unknown in Equation (12). Once the value of L is determined (either numerically or graphically) the future time course of $\bar{c}_t$ can be calculated while I, c_i, and λ remain (reasonably) constant.

The long-term nature of the $\bar{c}$ ($\bar{c}_f$) response can be predicted for the different possible values of L. If $L < 0$, $\bar{c}$ increases continually with t. If $L = 0$, from Equation (10) it follows that $\bar{c}$ increases linearly with time; $\bar{c} = \bar{c}_0 + t\,(Ic_i/z\theta_l)$. If $L > 0$, a final steady-state value of $\bar{c}$ ($\bar{c}_f$) will be reached. can be determined from Equation (10) by defining steady state as $d\bar{c}/dt = 0$, and so

$$\bar{c}_f = Ic_i/L\lambda \tag{13}$$

$\bar{c}$ will approach $\bar{c}_f$ asymptotically. Thus the most dramatic changes in $\bar{c}$ will be seen in the early times after any change in solute inputs.

The value of λ will vary with time, but not dramatically (Figure 4.3), unless water moves upward from a shallow water table in the root zone (Slavich and Yang, 1990; Thorburn et al., 1990b). In the absence of water tables, no significant errors are introduced by taking λ as constant, equal to its value at some time, say t_1 (Rose et al., 1979). λ will tend to increase or decrease with time to some asymptotic value. Therefore the error introduced in calculating L by taking λ at t, will be opposite in

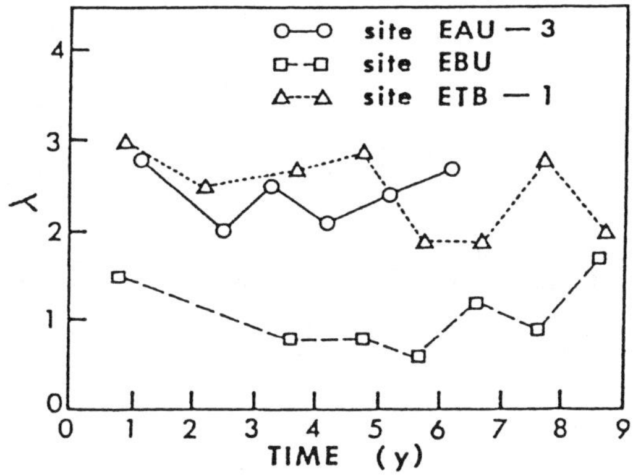

FIGURE 4.3. Variations in λ with time at three sites: (○), site EAU-3; (□), site EBU; (△), site ETB-1. (From Thorburn et al., 1990b. *Irrig. Sci.* 11:199–207. With permission.)

sign for $0 < t < t_1$, than for $t > t_1$. The value of λ will also be dependent on the profile hydrology. For example, $L < 0$ is interpreted as upward movement of water from a water table into the root zone, and under these conditions λ would be expected to be < 1. Similarly for $L > 0$, that is water moving downward below the root zone, λ would be expected to be > 1.

Measured $\bar{c}_t$ values have been compared with those predicted by the SODICS model at nine irrigated sites (Lyle et al., 1986; Thorburn et al., 1990b). Agreement between measured and predicted values has been good, except where shallow water tables developed in the root zone due to irrigation. Where there are shallow water tables in the root zone, the requirement for constant inputs of solute mass to the root-zone would not have been satisfied, and long-term predictions of $\bar{c}_t$ would be expected to be in error unless the variation in L and the upward contribution of salts into the root zone was quantified. Additionally, values of λ are highly variable in soils with high water tables. Slavich and Yang (1990) modified the definition of λ to allow for extreme variations in solute profile shapes that can occur in soils with shallow water tables.

Thorburn et al. (1990b) applied the model to 42 sites to assess the effect of irrigation management practices on soil salinity levels. They found that 19 of the sites were not in steady state, so application of common steady-state leaching fraction-based models [U.S. Salinity Laboratory (USSL), 1954] was invalid. Thorburn et al. (1990b) also examined the effect of "normal" variations in the input parameters to the SODICS model on model predictions (e.g., Figure 4.3). They defined "normal" as those variations that commonly arise in farm management, such as changes in cropping rotations or annual variations in irrigation application due to differences in rainfall. Generally these variations did not have a significant effect on calculated values of L or $\bar{c}_t$. Thus the model is applicable to many common field situations in irrigated areas.

Dowling et al. (1991) used SODICS, and a furrow advance-infiltration model, to examine the distribution of infiltration and leaching along furrows in an irrigated sodic duplex soil and a developing irrigation area. They found substantial leaching

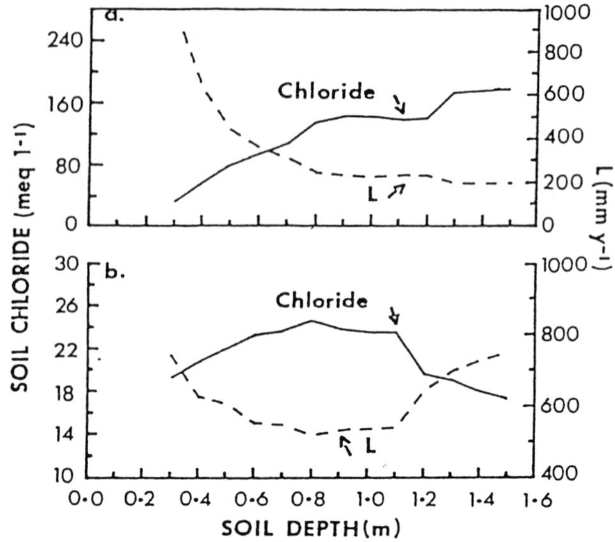

FIGURE 4.4. Variation in depth profiles of soil chloride and flux of water flowing through the soil matrix (*L*) at a site without (a) and with (b) preferential movement of water through soil cracks or pores. (From Thorburn and Rose, 1990. *Irrig. Sci.* 11:219–225. With permission.)

near the head ditch and where gypsum had been applied, and no significant leaching near the tail drain. These results had implications for irrigation management in the area, as there was concern over the development of shallow water tables due to irrigation. The normally slowly permeable nature of the sodic B horizon of this soil meant that it would have been extremely difficult to arrive at these conclusions using conventional soil physical investigations.

Preferential movement of water through cracks or large pores in soils will affect movement of solutes through the root zone. Slavich and Yang (1990) further modified the SODICS model to include a parameter accounting for the quantity of water preferentially moving through the profile. Thorburn and Rose (1990) used variations in L with depth (Figure 4.4) to calculate the flux of water bypassing the soil matrix in 36 irrigated profiles. They used SODICS to calculate L values, although any suitable model (e.g., Peck et al., 1981; Johnston, 1987) could be used (Thorburn and Rose, 1990).

The SODICS model has also been applied in a nonirrigated area, to assess the effect of land development on changes in L and $\bar{c}$ (Thorburn et al., 1991). Soil chloride profiles (to 2-m depth) had not reached steady-state conditions following clearing of shallow-rooted native *Acacia* forests at the site studied, so more commonly applied steady-state approaches (e.g., Edmunds et al., 1988; Dettinger, 1989) could not be used. Thorburn et al. (1991) found that after an initial "pulse" of water moved below the root zone immediately after clearing, water movement below the root zone was not significantly different under native forest, pastures, or crops. This initial "pulse" of water was not able to be identified by a the more common steady-state mass balance model.

Other transient approaches have been developed to estimate changes in L and leaching of solutes through the vadose zone following clearing of native forests. Allison and Hughes (1983) used the downward displacement of the soil solute concentration front (analogous to a wetting front) to calculate the amount of water that had moved below the root zone since clearing. This amount was divided by the time since clearing to give L. Their model assumes piston flow, and can be accurately applied only under these conditions and where clear solute concentration fronts can be identified. Walker et al. (1991) developed a generalized model based on solute front displacement that relieved these assumptions. The model requires profiles of solute concentration, θ, and Ψ at two times, to depths below the peak in solute concentration (possibly many meters). These data requirements may limit the application of the model where drilling for sampling is difficult, or drilling equipment is unavailable.

4.2.6 Considerations in Application of Transient Models

The solute transport models discussed in this section form a continuum of approaches, from the simplest, being mass balance approaches, to the most complex, based on the CDE. The spatial and temporal details of predictions increase with increasing model complexity. Increased complexity in models, however, also affects the data requirement of models, with a higher level of detail being required for input data to more complex models. From the perspective of the user, detailed data are not always available for input into solute transport models, and so as the complexity of a model increases so does the potential limitation to its practical application (Thorburn et al., 1990a,b).

Advisory agencies are commonly called upon to provide information on solute movement, but often have practical constraints on the amount and detail of data they can measure. This limitation affects the choice of models that can be used routinely over a wide range of soil types and managements. Also, if working in agricultural areas it is not always possible for land to be made available on commercial farms for experimental determination of model parameters. Because of these constraints, useful models are most likely to be practically oriented, as opposed to "research tools" (Addiscott and Wagenet, 1985).

Another consideration when choosing a model is the level of detail required in the prediction. Where it is not feasible to alter land management in the short term (as is the case in many agricultural areas), more detailed predictions given by more complex models may be of little practical benefit. Longer-term predictions given by simple mass balance models may provide useful predictions in this situation. However, these models require stable management through time and provide little detail of the depth distribution or short-term temporal variation of solute concentrations within the root zone. Where management practices vary with time, or short-term predictions are required, more detailed, CDE-based solute transport models need to be employed. Thus, a compromise may have to be made by model users when assessing solute movement in the vadose zone.

4.3 ONE-DIMENSIONAL STEADY-STATE SALT TRANSPORT IN THE VADOSE ZONE

4.3.1 Introduction

Steady-state salt mass balance is based on the principle of conservation of mass for a system in equilibrium. The mass of salt (volume of water by concentration) entering a system is in equilibrium with the mass of salt leaving the system. Once input or output parameters are changed by developments such as land clearing, cropping systems, irrigation, or water resources utilization, the system is no longer in equilibrium or steady state. The time to reestablish equilibrium may be on the order of centuries for landscapes or years for soil profiles. All systems are at best in dynamic equilibrium due to the cyclical nature of rainfall. The restraints of steady-state assumptions are reviewed in detail later in this section. Approaches for analyzing non-steady-state situations were treated in the previous section.

This section is concerned with applications of the steady-state salt mass balance to soil and land systems as a predictive tool in land management. It reviews models based on mass balance. Other types of models have been proposed, such as those based on salt displacement, in which salt is displaced from the soil surface in a manner similar to equilibrium chromatographic adsorption as discussed by Bolt (1979). This approach will not be treated further.

4.3.2 Leaching Fraction

The USSL (1954) introduced the steady-state salt mass balance concept to look at the leaching of salts in soils. The steady-state model assumes that an equilibrium occurs between the inputs and the outputs of salt after a given period where the change in salt storage becomes zero. The USSL used the term *leaching requirement* as the fraction of applied water that needs to be drained through the root zone to control soil salinity. This equation can be expressed as

$$LR = \frac{c_i}{c_o} = \frac{Q_o}{Q_i} \tag{14}$$

where LR is the leaching requirement, Q_i is the quantity of water entering the soil, usually expressed as a depth (mm), Q_o is the quantity of water draining below the root zone, c_i is the concentration of electrolyte in the input water, and c_o is the concentration of electrolyte in the water draining below the root zone.

It should be noted that c_i and c_o are commonly represented by a corresponding electrical conductivity (EC), to which they are proportioned.

This concept has probably been the most widely applied model of salt movement. There have been a range of variations and extensions introduced, and some relaxation of the conditions of steady state allowing its application in a wider range of situations. Scofield (1924) appears to have been the first to use this concept of a mass of salt moving in steady state (but not in algebraic form) when comparing the mass of salt in irrigation and drainage waters.

Lyle et al. (1986) compared three models to predict solute leaching under irrigation with a range of irrigation water salinities on a sodic soil in southern Australia.

They compared the steady-state salt balance model of the USSL (1954) with the non-steady-state model of Rose et al. (1979), for slowly permeable soils, with a summation of the salt mass in the profile over a number of small time steps to account for seasonal variability. The steady-state mass balance model gave as good a prediction of leaching flux as the other models. The non-steady-state model also allowed a prediction of the time period to reach steady state and the steady-state salinity value.

4.3.3 Extensions of Steady-State Mass Balance Models

4.3.3.1 Leaching of Salt-Affected Soils

Panin (1962) was interested in predicting the effectiveness of leaching in removing salts from saline soils and, after reviewing earlier Russian work, developed a model based on steady-state concepts but with a leaching efficiency term. Panin found that the salt content of the soil differed from the leachate concentration. This presumably was due to both the nonattainment of steady state in such short time periods after significant hydrologic change and also to the influence of macropore flow as discussed later in Section 4.3.5.6.

The equation is of the form

$$D_i = Q_F \mathrm{Ln} \frac{s_i}{s_o} \tag{15}$$

where s_i is the total initial salt content in the soil, s_o is the salinity of the soil after leaching, D_i is the volume or equivalent ponded depth of applied water for leaching, and Q_F is the "leaching constant" that accounts for the soil properties determining the release of salts, in the same units as D_i.

Boumans (1983) also was concerned with leaching of salts from saline clay soils in Iraq. Boumans included a salt contribution from a shallow groundwater source in the steady-state salt balance equation and a leaching efficiency term, f, which related the salt concentration in the soil water at a moisture content close to field capacity to the concentration of the water actually draining below the nominated soil layer. Boumans also corrected for the salinity difference due to the technique of measuring the salt content on a 1:1 soil water extract compared to the field moisture content. The leaching efficiency term had a value of around 0.6, which is again probably typical of a system not yet in steady state and influenced by macropore flow.

4.3.3.2 Estimation of Ground-Water Recharge from Salt Profiles

One of the earliest approaches to estimating recharge from salt (chloride) balances was presented by Eriksson and Khunakasem (1969). The authors compared the chloride concentration in rainfall with the chloride concentration in groundwater under the Israel coastal plain to evaluate recharge. They integrated the rainfall-deposited chloride in the groundwater along the groundwater flow lines to the sea and concluded the recharge estimates were reasonable when compared to more conventional water balance estimates.

Because the recharge estimates were relatively high, on the order of 20 to 300 mm year^{-1}, it was assumed that any historic or evaporite salts in the aquifer would have been leached. In situations where there is a source of chloride from rock weathering, the estimates will be in error. Another problem with using aquifer salt contents is that the time constant for the whole system to come to equilibrium is very long, and recent changes in land management such as clearing may mean the aquifer could take centuries to respond. Thus steady-state conditions may not be achieved. Peck and Hurle (1973) calculated the time periods for a number of small catchments in western Australia to respond to clearing based on salt export from the catchments. Their time periods ranged from 20 to 400 years.

Allison and Hughes (1978) compared estimates of recharge from tritium concentrations in soil water with the steady-state mass balance of chloride. The soil chloride below a depth of 2 m was calculated as a concentration in the soil solution at field water content. The agreement between methods was good and the authors concluded that chloride could profitably be used as a tracer in hydrologic studies provided the history of land use was known.

Allison et al. (1985) applied the technique to a semiarid area of southern Australia. They expressed the recharge in terms of the chloride accession in rainfall to the chloride concentration of soil water in the soil profile. This was calculated at the volumetric soil water content at sampling and expressed as milligrams chloride per cubic meter of soil water. The soil depth chosen was below the soil profile chloride bulge at a depth of around 22 m. The selection of this depth was based on the ^{18}O analysis that suggested wetting to a depth of 2.5 m in recent times. They also used an alternative technique of cumulative chloride concentration below a depth of 5 m compared to the average annual chloride accession in rainfall. A comparison with recharge estimates based on tritium profiles indicated a very close agreement.

Sharma et al. (1983) expanded the mass balance to include a change in the salt storage term. This allows the extrapolation to situations of zero, downward, as well as upward recharge (from a shallow water table). This approach has been discussed in the section on transient models.

Edmunds et al. (1988) added components of dryfall, precipitation, and adsorption to the mass balance equation but considered only conditions in steady state, where these processes were effectively zero. They assessed sites in Cyprus and Sudan against tritium methods and found satisfactory agreement at rainfalls between 180 and 410 mm year^{-1}. The chloride technique overcame some of the technique and sampling problems common with tritium. It also gave a long-term average recharge estimate.

In a manner similar to that of Peck and Hurle (1973), Dettinger (1989) applied the steady-state model to estimate recharge to a number of basins in the western United States using the chloride concentration of rainfall and groundwater, concluding that the chloride mass balance was a useful reconnaissance method to estimate groundwater recharge.

The alternative to salt balance estimates of recharge is direct measurement of water flux. As pointed out by Scotter et al. (1988), the problems of using measured soil hydraulic properties to estimate drainage by Darcy's law are major. The need to know the $K - \theta$ relationship at water contents less than saturation introduces large measurement errors. Soil water transport models such as those of Campbell (1974) require parameters that are difficult to measure at less than saturation, have high spatial variability, and suffer hysteresis. Allison (1988) also comments on the difficulties of the water flow method and considers that estimates by this method can only be considered "rough" due to the inaccuracies in the measurement of unsaturated soil properties.

4.3.3.3 Estimation of Root Zone Salt Content

Hoffman and van Genuchten (1983) combined the continuity equation for one-dimensional water flow with the steady-state mass balance for salt to calculate the slope of the salinity concentration gradient with depth. This was used to calculate water uptake-weighting functions to estimate root zone salinity and relate to plant salt tolerance yield relationships. Smith and Hancock (1986) derived a more generalized relationship that is independent of any nominated water uptake pattern. Their equation can be expressed as

$$C_w = c_i \left\{ \mathrm{Ln}\left(D_d/D_i\right) / \left[\left(D_d/D_i\right) - 1\right] \right\} \tag{16}$$

where C_w is the weighted salt content over the nominated root zone depth, D_d is the depth of water draining below the root zone, D_i is the depth of applied water, so that D_d/D_i can be replaced with the leaching requirement (LR) at the bottom of the root zone to maintain the root zone salinity below a threshold value, and C_w can be replaced with C_w^*, the appropriate salt tolerance rating for the particular crop.

The question of average vs. water uptake-weighted root zone salinity has been discussed in the literature, for example, Shalhevet and Bernstein (1968), Bernstein and Francois (1973), and Devitt et al. (1984). Plant salt tolerance experiments have generally been carried out in containers under high leaching rates to obtain relatively uniform salt profiles with depth. In natural soils in semiarid areas, particularly with medium to high clay contents, a very nonuniform salt profile results from salt accumulation and reduced plant water uptake with depth, partly due to insufficient wetting. This nonuniform salt profile with depth is probably much better represented by a water uptake-weighted salinity value to relate to the experimental conditions in which plant salt tolerance is assessed (Shaw, 1988a).

4.3.3.4 Irrigation Water Quality Assessment

Bernstein (1967) attempted to improve on the irrigation water quality suitability assessment by incorporating the significant management factors into the steady-state salt balance equation. The author identified these as soil, climate, and water management, modifying Q_o of Equation (14) (equivalent to D_d) to

$$D_d = D_i - D_e$$

where D_e is the depth of water evapotranspired. D_i is then expressed as a function of the infiltration rate and the time of irrigation with D_e expressed as evapotranspiration rate over the time of the irrigation cycle. Bernstein was able to reduce leaching fraction, LF, to two limiting cases:

1. Drainage rate is limiting:

$$LF = \frac{D_r}{E + D_r} \tag{17}$$

where D_r is the daily drainage rate over the time period considered, and E is the daily evapotranspiration rate over the same time period.

2. Drainage rate is nonlimiting:

$$LF = 1 - \frac{E\,t_i}{I\,t_i} \tag{18}$$

where t_i is the time period of infiltration over the irrigation cycle, and I is the infiltration rate over the same period.

From the leaching fraction, root zone salinity can be determined. The difficulty is in determining the rates of evapotranspiration and infiltration, which are far from constant over the period of irrigation.

4.3.3.5 Prediction of Leaching Fraction

Because the steady-state salt balance of a soil profile is inextricably linked to the soil water balance, then the salt balance and readily measured soil properties associated with soil hydrology should be able to be used to predict deep drainage and leaching fraction under a range of water managements.

Shaw and Thorburn (1985a) examined over 700 soil profiles in northeastern Australia and developed a physically based empirical model relating key soil properties and rainfall to salt content at the bottom of the root zone and thus soil leaching fraction and deep drainage. Soil clay content, cation-exchange capacity (CEC) of the clay fraction, expressed as CCR (CEC-to-clay ratio, an index of clay mineralogy), exchangeable sodium percentage (ESP), and rainfall were highly significant variables explaining around 70 to 80% of the variability in soil salt content at a depth of 0.9 m. By including rainfall over a range from 200 to 2000 mm, a way of predicting the deep drainage (D_d) under changed water managements was possible. Shaw and Thorburn (1985a) and Shaw et al. (1987) used this model to predict leaching fraction under irrigation satisfactorily, with an adjustment for the change in electrolyte concentration of the irrigation waters. The significance of two soil properties, exchangeable sodium percentage and clay content, on predicted leaching fraction is illustrated in Figure 4.5.

The general form of their relationship for specified clay content and CCR groups was

$$\frac{c_i}{c_o} = \frac{a\,R_m}{ESP^k} + b \tag{19}$$

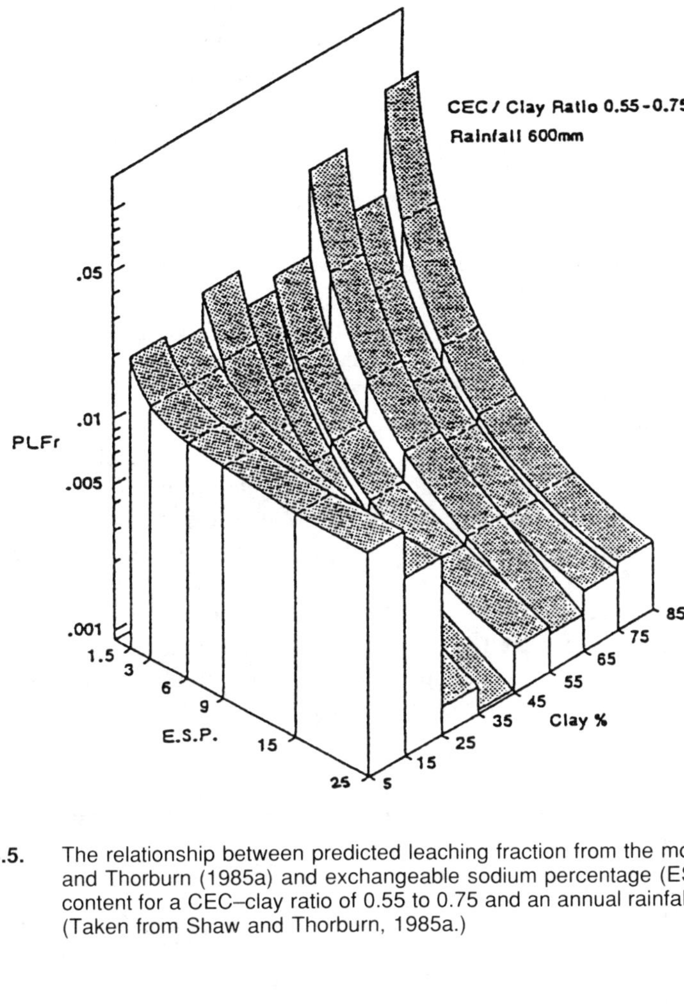

FIGURE 4.5. The relationship between predicted leaching fraction from the model of Shaw and Thorburn (1985a) and exchangeable sodium percentage (ESP) and clay content for a CEC–clay ratio of 0.55 to 0.75 and an annual rainfall of 600 mm. (Taken from Shaw and Thorburn, 1985a.)

where a and b are regression coefficients, R_m is the depth of median annual rainfall, ESP is the exchangeable sodium percentage at the bottom of the root zone, and k is a coefficient that linearizes the influence of ESP.

By considering salt content below the active root zone depth, runoff and evapotranspiration are empirically accounted for within the coefficients of the regression equations. The Shaw and Thorburn model [Equation (19)] is based on a summer dominant rainfall regime and will need to be adjusted for the rainfall effectiveness of other climatic situations as indicated by Yaalon (1983). The prediction of leaching fraction from soil properties allowed a more quantitative irrigation water-quality assessment model (Shaw and Thorburn, 1985b), which is particularly appropriate to more slowly permeable soils in which soil properties rather than water management determine leaching fraction. Heavy clay soils with infiltration rates between 2 and 20 mm day^{-1} are commonly irrigated in Australia, Africa, and India. These rates are considerably less than the recommended minimum rates of 25 mm day^{-1} (USSL, 1954) and 60 mm day^{-1} [U.S. Bureau of Reclamation (USBR), 1982].

Predicted D_d using a rainfall of 800 mm was related to ponded infiltration rate, IR_p, using the soil analytical data available for a range of clay soils where IR_p was measured. This is shown in Figure 4.6 to illustrate the broad relationship with saturated soil hydraulic conductivity. Shaw and Yule (1978) and Gardner and Coughlan (1982) measured ponded infiltration rates after 3 to 7 days in field bays 5 × 5 m, with sheet steel walls inserted to a depth of 1 m to minimize lateral water flow through cracks. The measurements were made on soils representing Vertisols and Alfisols (Soil Survey Staff, 1975) ranging in clay content from 23 to 83%, and with an exchangeable sodium percentage at 0.9 m of 0.9 to 39%.

Electrical conductivity was used instead of chloride in the Shaw and Thorburn model because the degree of leaching of a soil is related to the total electrolyte content (Frenkel et al., 1978), and the other ions deposited in rainfall, particularly sulfate, make a contribution to the electrolyte. Matthias et al. (1986) also used electrical conductivity instead of chloride due to ease of measurement and because there was a significant relationship with chloride. The salt content of a soil at field water content around "field capacity" was used instead of the content at saturation as that would be the water content at which downward flow of water and thus leaching would become very small. The shape of the relationship between the soil salt content at the bottom of the root zone (0.9 m) and ponded infiltration rate is given in Figure 4.7, indicating the significance of soil hydraulic properties on soil salt content.

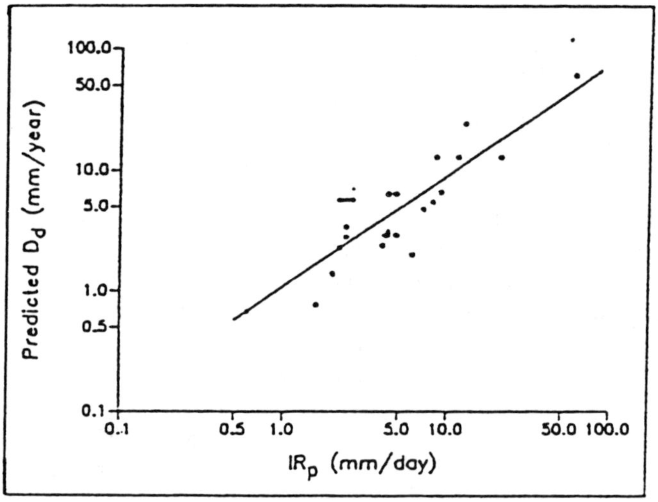

FIGURE 4.6. Relationship between predicted deep drainage (D_d) for a rainfall of 800 mm $year^{-1}$ from the model of Shaw and Thorburn (1985a) and measured ponded infiltration rate (IR_p) after 3 to 7 days for a range of soils in northeast Australia. (From Shaw, 1988b. *Soil Use Manage.* 4:120–123. With permission.)

4.3.4 Accuracy of Recharge Estimates Using Salt Balance Methods

Measurement and validation of the steady-state model can be quite difficult because of the very small drainage rates that often occur in semiarid and arid situations and the major problem of spatial variability. Third, in high leaching situations, the soil salt content is so low that significant measurement errors can result. Alternatives are direct measurement of solute flux, comparison with tritium tracers, or water balance approaches. There are significant problems with these alternative approaches also. Techniques for solute flux measurement are point measurements and are affected by macropore flow. Wagenet (1986) considered that there was no suitable direct measurement on a field basis. Measurement accuracy is a real problem at low flow rates of less than 1 mm day^{-1}. The errors in using a water balance technique with small deep drainage (D_d) values are also major in that a small value is derived from the subtraction of two much larger numbers, water input and evapotranspiration. Gee and Hillel (1988) review the various methods of estimating groundwater recharge (D_d) for arid regions in contrast to humid areas. While the limitations of all the methods are considerable, particularly water balance methods, the authors were more positive about both natural and introduced tracer techniques.

Sukhija et al. (1988) attempted to estimate the errors associated with recharge predictions using the steady-state salt balance method. They suggested that the noninclusion of dry fallout would contribute less than 10% error. The authors sampled five sites twice, 9 months apart, and were convinced steady-state conditions applied. The variation in mean and total chloride content was within 30%. They concluded that the total error in estimating recharge (including some small errors in water and salt content measurement techniques) would be around 30 to 40%.

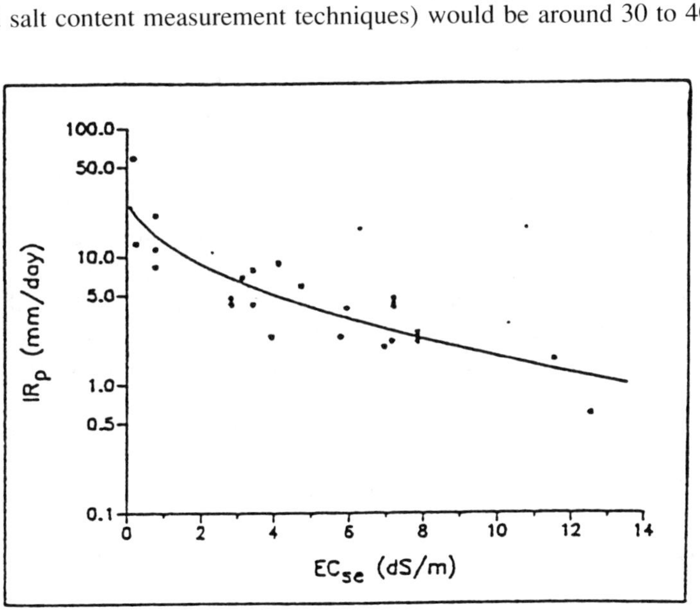

FIGURE 4.7. Relationship between ponded infiltration rate (IR_p) after 3 to 7 days and soil salt content (EC_{se}) at 0.9 m for a range of soils in northeastern Australia. (From Shaw, 1988b. *Soil Use Manage.* 4:120–123. With permission.)

4.3.5 Assumptions in the Use of Steady-State Salt Balance Models

The application of the steady-state mass balance model requires a number of assumptions. These include the uniform application of water to the area, that salt is conserved, and no precipitation or plant uptake of soluble salts occurs as outlined by the USSL (1954). Where the model is linked to the water balance, which is the general case, a range of additional assumptions need to be met. The validity of these assumptions is discussed below in some detail to illustrate the potential errors associated with the application of the steady-state model in field situations.

Edmunds et al. (1988) give a detailed account of the basic assumptions in using tracer or solute-based methods for recharge estimations.

4.3.5.1 Steady-State Conditions and Minimum Time Steps

Inputs of rainfall are strongly episodic and for arid and semiarid environments, where evapotranspiration is in excess of rainfall on an annual basis, significant soil profile wetting will occur only over several days of the year, which can hardly be called steady state. However, outputs through soil water redistribution and drainage will occur over several weeks. Over periods of decades the averaging of inputs and outputs will have only a small effect. Provided the contact time for the infiltrated water with the matrix is sufficiently long for some quasi-steady-state soil solution equilibrium to occur, the salt content should reflect the steady-state hydrology.

Johnston (1983) used steady-state conditions to evaluate recharge based on soil chloride profiles with a minimum time step of 1 year. Kirkby (1985) quotes evidence for silicate chemistry to come to equilibrium in the soil solution within 100 h of contact. It is expected that wetting of highly sodic aggregates will take time in the order of days, which should still be within the time frame of soil water redistribution events. This is self-compensating to some extent in that sodic soils have reduced vertical permeability and hence water will remain in the profile longer. Kirkby (1985) gives predicted chemical weathering changes in various soil parent materials and indicates that reasonable degrees of weathering may take in the order of 80,000 years. Sears and Langmuir (1982) in considering the chemistry of soil water found ion exchange accompanied by selective adsorption were the dominant controls on the chemistry of water in soil solution suction extractors. While this technique has been severely criticized for carbon dioxide degassing errors (Suarez, 1986), the authors quote evidence of these sorption and exchange reactions being complete within 48 h, while calcium carbonate and clay mineral reactions require time periods in the order of months to years.

Otoma and Kuboi (1988) modeled the time to come to equilibrium for profiles with 15 tonnes ha^{-1} of sewage applied annually to the surface. The soils contained less than 8% clay and under annual rainfalls of 1200 and 1800 mm the top 1 m of the soil profile was predicted to take 10 and 7 years, respectively, to establish a new chloride equilibrium. Wright (1928) measured monthly changes in surface soil salinity, and found considerable differences between the end of the dry season and the end of the following wet season, which is not unexpected for surface soils. For sampling of the soil matrix at or below the bottom of the root zone, the changes will not be as significant. However, the effect of particularly wet periods on salt concentrations causes greater variation. Talbot and Bruce (1974) found significant leaching of salts out of an irrigated clay alluvial soil to a depth of 0.6 m in the Lockyer Valley in southeast Queensland after a rainfall of 768 mm (the normal annual rainfall) over a 3-month period. The soil had been irrigated with a water of electri-cal conductivity 6 dS m^{-1} with a relatively low sodium adsorption ratio of 4.5. Shaw and Yule (1978) found a significant movement of salt out of the surface 0.9 m of a range of clay soils after a very heavy rainfall period where runoff was restricted. While these events will be less significant at lower salinity levels and over the long time periods being considered, they do have an important influence in the time of sampling of soil profiles for use in salt balance models or for validation purposes. An example of rainfall variation for an area of northeastern Australia is given in Figure 4.8 to illus-trate the importance of considering time of sampling. In fact, it is hard to consider any steady-state assumption applicable in such a variable rainfall regime.

The appropriate use of the steady-state model requires a matching of the time constant of change in land use resulting in a change in salt output. The time constant varies considerably depending on the scale and the rate of recharge. Peck and Hurle (1973) estimated between 20 and 400 years for catchments of less than 100 ha, while Allison et al. (1985) estimated between 16,000 and 30,000 years for deep (around 20 to 30 m) soil profiles at a recharge rate of about 0.1 mm $year^{-1}$. Thorburn et al. (1990b) calculated equilibrium times generally between 3 and 50 years for a depth of 0.9 m in irrigated soils. The time period reflects the interaction between soil properties determining soil water movement and the quantity and quality of irrigation water. Edmunds et al. (1988) calculated that for a 25-m soil profile at an average recharge rate of 1 mm $year^{-1}$, the accumulated salt represented a period of some 2000 years. Thus it is important to ensure land management changes are reasonably constant over the period involved, particularly for situations of small annual recharge under dryland conditions.

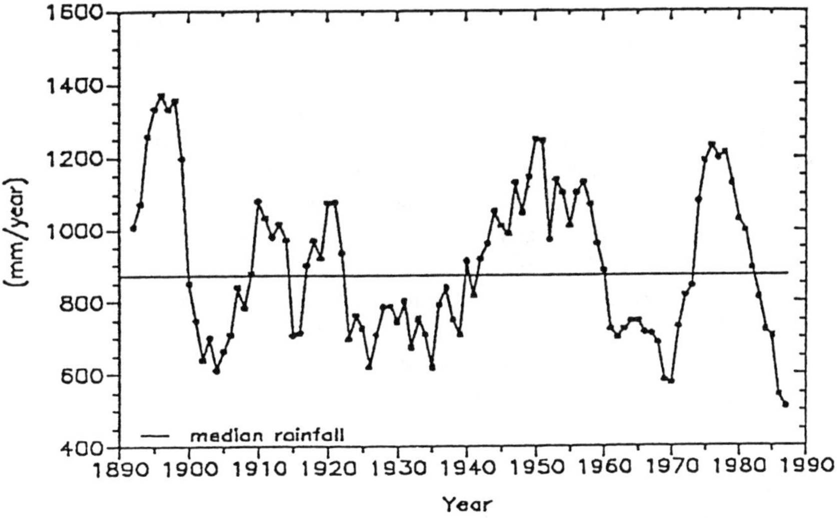

FIGURE 4.8. Five-year moving average rainfall for Clare, northeastern Australia, illustrating a large cyclical variation.

Jury and Pratt (1980) commented that where a change in salt composition occurs, cation exchange and precipitation reactions will occur and delay the time to equilibrium. This can also occur through salt mobilization after a change in hydrologic balance through land development such as land clearing or irrigation. This is less of a problem with chloride ion, which does not undergo transformations in the soil.

Wierenga (1977) and Beese and Wierenga (1980) compared the relative merits of simple steady-state and transient models of solute movement where strict steady-state conditions were not met and found the more complex transient state models were generally unnecessary even for intermittently irrigated fields. Their basis was that solute movement is a function primarily of the cumulative amount of water flowing past a given depth and cumulative water could be used instead of time, as the necessary input data are often unavailable for transient models.

4.3.5.2 Constancy of Salt Inputs

Assumptions of constant salt input over long time periods are open to question. Allison et al. (1985) examined cumulative chloride vs. cumulative water content for soil profiles deeper than 5m and concluded for southern Australia that the rate of recharge was fairly constant over the last 13,000 to 16,000 years, with only a small change for the preceding 14,000 years.

For rainfall salt inputs there is the known change in salt concentration with distance from the coast: for example, as described by Probert (1976), Hutton and Leslie (1958), Isbell et al. (1983), and Eriksson and Khunakasem (1969). There is considerable variation in the concentration of salts during a rainfall event and over a season; for example, Reynolds and Pomeroy (1988) give variations in bulk precipitation of between <0.5 to 18 mg L^{-1}. Allison and Hughes (1978) determined from data collected by Hutton that the annual flux is approximately constant because low annual rainfalls tend to have higher salt concentrations than larger annual rainfalls. The chloride in rainfall data collected by Hutton and Leslie (1958) and earlier published data they reviewed suggests a significant decrease in chloride concentration with distance from the coast up to a distance of around 150 km.

Dryfall and vegetation interception as well as use of fertilizers can affect salt inputs, which may modify the predicted result. However, the effects are usually a small percentage of a small quantity and can essentially be dismissed over the time frame necessary for the appropriate application of the model. Dust in rainfall collectors can overestimate rainfall accessions because it normally cycles within a catchment but is trapped in collectors (Edmunds et al., 1988). Møberg et al. (1991) collected around 580 kg ha^{-1} of dustfall over a 5-month period in northern Nigeria.

The accurate measurement of quantity of rainfall is a problem and this is particularly so on landforms with significant slopes. Thus the overall uncertainty in rainfall salt inputs needs to be considered.

4.3.5.3 Nonreactive and Reactive Salts

The assumption of no salt precipitation in the application of the model poses considerable problems. Traditionally chloride has been used because it is always conserved and not transformed. Problems arise in high charge density clay soils where anion exclusion occurs. This results in repulsion of anions from the clay particles, a lower salt content, and an overestimate of the actual water movement through the matrix (e.g., Bond and Smiles 1988; Appelt et al., 1975). Travel times for the chloride ion were found to be faster than for tritium tracers by Weirenga and van Genutchen (1989) in a sandy soil due to anion exclusion. Sulfate is an important anion also deposited in rainfall but which undergoes transformation, or more often precipitation with calcium as gypsum, particularly in arid and semiarid regions. The problem here is that gypsum is partially soluble, as are other similarly behaved salts that precipitate from the dissolved CO_2 in rainfall, such as sodium bicarbonate and sodium carbonate. These salts influence soil hydraulic conductivity and thus have a direct effect on salt leaching. The problems accounting for gypsum precipitation in irrigation situations have been well illustrated by Willardson and Hanks (1976). An alternative method used by Shaw (1988a) was to use electrical conductivity as the basic measurement and to correct to a more concentrated water content by accounting for the nonchloride composition of the total salt content. This is a simpler method requiring far fewer data than salt concentration models proposed in the literature that are more exact but which require input parameters that are often not available.

4.3.5.4 Change in Soil Water Storage Can Be Neglected

The maximum range in soil water storage expected for the root zone of clay soils with high water holding capacity is around 180 mm based on the measurements of Gardner and Coughlan (1982). McCown (1971) found values of 180 mm for soils from north Queensland. Over a 10-year period, with an annual rainfall of, say, 600 mm, that is 6000 mm for the period, and assuming accounting periods that are not at the extremes of the soil moisture range, then this error will be very small and can be neglected. Allison (1988) also considered that changes in soil water storage could be neglected if calculations were made on an annual basis.

4.3.5.5 Neglect of Runoff

Runoff is often less than 10% of annual rainfall. Lawrence and Sinclair (1989) presented data for runoff over a 3-year period from 1985 to 1989 under native Brigalow scrub and cropped areas in central Queensland. Maximum runoff occurred on the cropped catchment at 6% of annual rainfall compared to 3% under Brigalow. The Snowy Mountains Engineering Corporation (SMEC, 1988) gave runoff figures for a range of selected catchments in northern Australia over the period 1921 to 1981. With the exception of one catchment, these figures are all less than 10% of annual rainfall. This illustrates that a bias will be introduced, with higher rainfall regions having proportionally smaller effective rainfalls. These figures are determined by soil type, land management, vegetation, and topography. It is possible to broadly correct to some measure of effective rainfall but the error of correction may be as inaccurate as the individual site differences, and thus the assumption of no runoff will be equally valid.

For soils with coarse textures and high infiltration rates, runoff will be a small component of the water balance, depending on rainfall intensity. Effectively this means that a soil type bias will also be introduced into this assumption.

4.3.5.6 Macropores and Preferential Flow

Problems arise where soils have structural development and macropores occur. Macropores and preferential flow paths are well recognized as major conduction pathways in soils (Thomas and Phillips, 1979; Bouma et al., 1981; Steenhuis et al., 1988). Depending on the pore size, these macropores will tend to be active only when the soil is at or near to saturation (Seyfried and Rao, 1987). Bouma et al. (1981) found that macropore flow was significant as soon as the ponding capacity of the soil surface was exceeded. Kneale and White (1984) showed that macropore flow was proportional to rainfall intensity. Macropores will thus conduct rainfall through the matrix quite quickly until the soil water potential falls to a point below which the macropores cannot continue to conduct water except as films at a rate closer to that of the soil matrix. A considerable proportion of the research on preferential flow has been directed to the evaluation of loss of surface-applied nutrients such as nitrate (White et al., 1983), groundwater pollution or groundwater recharge and water table response (Burch et al., 1987), and drainage design (Steenhuis et al., 1988).

Sharma (1988) evaluated soil profile chloride and groundwater chloride levels as estimates of recharge in western Australia. By taking the soil chloride contents below the root zone (here >10 m), and the chloride content of groundwater, Sharma concluded that over 50% of recharge was through macropores in structured lateritic soils and even sands. The rainfall pattern where this work was done is a strong Mediterranean climate with wet winters, where the frequency of soil saturation at the surface would be expected to be much higher than for a summer rainfall regime of more frequent, high-intensity storms with associated higher runoff.

White et al. (1986a) studied leaching of herbicides through undisturbed soil cores (with clay contents between 35 and 52%) and found continuous water application resulted in greater leaching, while intermittent application gave less leaching, which they attributed to diffusion of the herbicide into the soil aggregates. They suggested that preferential flow led to short travel times for soils with an associated relatively low volume of water. Thus a considerable proportion of the applied water resulted in wetting of the matrix. Jarvis et al. (1987) examined the rainfall intensity data for soils at Silsoe, United Kingdom and determined that nearly 70% of rainfall infiltrated the clay matrix. This is in contrast to the high volumes of water transmitted through macropores as found by Peck et al. (1981) and Watson and Luxmoore (1986). Thus the degree of bypass flow is probably related to the size range of soil structural units and the time period the soil surface remains saturated. The net effect will cause errors in the input term by bypassing input below the root zone in situations where the soil water content is frequently above "field capacity" and close to saturation, or saturated for extended periods.

Van Genuchten and Dalton (1986) reviewed a number of models for coping with solute movement in aggregated soils containing macropores. The models assume solutes are moved through well-defined pores or cracks with a known geometry. Salt diffusion equations were used to describe solute movement within the smaller pores of the matrix. These models require greater data inputs and are discussed in the following sections on transient models. Thorburn and Rose (1990) assessed bypass flow from the shape of the soil salt profile. Addiscott et al. (1986) developed a simplified layer-based model consisting of immobile and mobile phases through which solute moved to equalize the concentration gradients.

Jury (1983) considered the nonuniform water and solute transport even in uniform sand under uniform water application to be a major problem facing the application of various models to the field. His transfer function model was an attempt to account for the range of pore water velocity variations found in the field. This approach is also discussed in Section 4.2 on transient models.

Macropores reduce the quantity of rainfall or irrigation available for movement through the matrix by bypassing it through macropores. For recharge estimates, the matrix salt content will significantly underestimate the actual recharge. Because rainfall following irrigation is a major cause of significant deep drainage through macropores where present, weighting for average salt concentration in the input will be more in error than for uniform soils. In dryland situations, because macropores bypass some of the salt input beyond the root zone, the salt mass balance errors will tend to cancel out although groundwater recharge will be underestimated.

A consequence of macropore flow can be the existence of strong salt gradients away from the macropores. Johnston et al. (1983) looked at the changes in Rhodamine WT dye with radial distance from a large flow path in a clay subsoil. They found that the distance to where the concentration dropped to half the initial was about 150 mm, which was double the prediction for the diffusion process alone and indicated that bulk movement by sorption was also occurring. A distance of 150 mm would suggest that for many soils quite significant matrix wetting will still result from macropore flow.

Allison (1988) quotes Zimmermann et al. (1967), who concluded that because the distance between large and small pores is small, the concentration of a tracer will become equal in all pores and thus the effect of macropores will not be great.

A second problem is fingering, which occurs in layered soils where finer textures overlie coarser-textured subsoils resulting in a lower hydraulic conductivity material overlying a higher conductivity material. An unstable wetting front develops and the flow breaks into separate flow streams called fingers as described by Hillel and Baker (1988) and Starr et al. (1986), among others. This will also result in bypassing of the subsoil matrix.

The problems associated with bypass flow attributed to both of these conditions can be detected from soil profile salt contents: in the case of macropores, by a reduction in soil matrix salt content with depth where sufficient depth of sampling is available, and in the case of fingering, by a change in clay content.

Thus preferential flow will mean that water moving in macropores under rainfall is not in equilibrium with soil matrix salinity. At the water content where downward leaching has effectively ceased, that is, a soil matric potential of around –0.1 to –0.3 bars, preferential flow will have long ceased.

White (1985c), in a review of the influence of macropores on transport through soil, concluded that "whether macropores increased or decreased the amount of solute leached depended on several factors, namely:

- The location of the solute, whether at the soil surface, on the surface of peds, or within peds
- The ratio of macropore to matrix flow
- The saturated hydraulic conductivity of the matrix
- The antecedent water content of the soil
- The contact area between the bypass flow and the relatively static matrix water
- The rate of solute diffusion between the mobile and immobile water volumes in the soil"

White concluded that more research is required on this aspect, particularly because convective dispersion-based models also have great difficulty in field soils with macropores.

4.3.5.7 Soil Matrix Salinity Equivalent to c_0 of Equation (14)

Shaw (1988a) considered salt profile shape in soils in terms of the steady-state salt balance applied to increasing depth increments in a soil profile. Assuming the salt uptake by vegetation is small, then from the conservation of mass, salt concentration will increase with depth in a profile (without a shallow water table) until the effective removal of water by plant roots is zero. From this point downward for a uniform soil, the salt concentration should remain constant with depth, this salt concentration being in equilibrium with the salinity of the water draining below the root zone of the particular soil. Explanations of variations in salt profile shape from this ideal shape (zero flux plane) are explained by Shaw (1988a) and Allison (1988).

The interpretation of nonstandard profile shapes is open to debate. Allison (1988) illustrated three profiles, as shown in Figure 4.9, and interpreted them as given in the figure caption. Gardner (1967) used alternative interpretations, as did Shaw (1988a). Profile c of Allison (1988) (Figure 4.9c) would suggest some bypass flow, as does profile b (Figure 4.9b), or the effect of both upward and downward fluxes of water movement if the depth to the water table is less than 3 m. For deeper depths the profile shapes probably reflect a fluctuating water table and historic changes in recharge rates. The depth of sampling needs to be considered in relation to the depth of major water uptake by vegetation, that is, the active root zone depth, and the influence of a water table, in order to sample the most appropriate "steady-state" profile below the root zone.

4.3.5.8 Estimation of Soil Matrix Salt Content

A further problem is the analysis technique for soil matrix salt content. Practical methods require water extraction involving the drying, grinding, and mixing of a soil sample. This is quite atypical of the more likely soil composition on the outer regions of aggregates and macropores, where drainage water will be in equilibrium. The problem will be related to the rate of diffusion of salts within aggregates and the degree of wetting of aggregates. Hantschel et al. (1988) looked at the differences in extracts of mixed soil and leaching through undisturbed cores of the surface 60 mm of a range of permeable soils and found significant differences in cation composition. Hamid and Warkentin (1973) studied the leaching of salts out of various-sized aggregates and found effectiveness of leaching increased with decreasing diameter

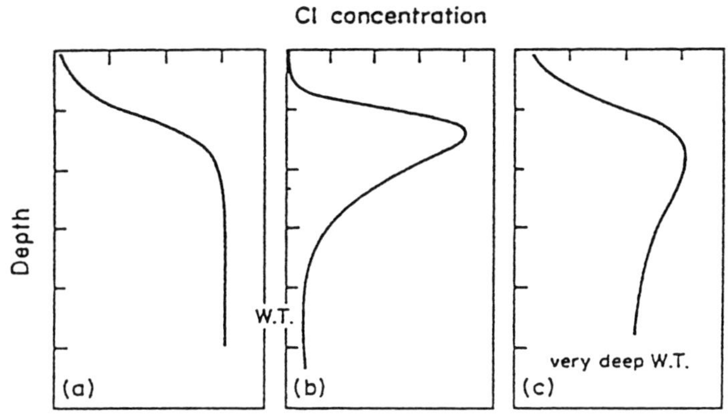

FIGURE 4.9. Schematic chloride profiles of soil water: (a) piston flow with abstraction of water by plant roots; (b) abstraction of water by plant roots but with either preferred flow of water beneath the root zone or diffusive loss of salt to the water table (W.T.); (c) a profile that may reflect the recharge history of a site. (Taken from Allison, 1988.)

of aggregates. In sodic soils, salts can be trapped within aggregates (Blackmore, 1976). The release of these salts in conventional water extract soil analyses may give higher salt contents than those actually draining through the soil profile. This problem can also occur if enhanced recharge removes salt that used to maintain the surface of aggregates in a flocculated condition. This process then traps salt within aggregates as described by Gvirtzman et al. (1988). Because the diffusion process is slower through dispersed soils, this will again result in a bias in the results toward high measured soil salt contents for higher sodicity subsoils with larger aggregate sizes, where soil dispersion of the outer layers of soil aggregates will reduce water movement and diffusion.

Lyle et al. (1986) calculated matrix salinity as the product of soil chloride content times bulk density divided by water content as an equivalent to leachate chloride content. Shaw and Thorburn (1985a) and Shaw (1988b) used the salt content of a soil-water suspension corrected to field water content using methods based on soil-to-water ratio and salt composition (Shaw et al., 1987). Others have used solution extractors and others calculated to the water content at sampling. No method is exact. Where a water content is drier than field maximum water content, bias is introduced by sampling at a water content at which no effective water movement is occurring.

4.3.5.9 Rainfall Inputs and the Soil Salt Content below the Active Root Zone

Shaw et al. (1987) calculated the rate of release of salts from weathering of basalt or granite rocks under an annual rainfall of 800 mm at the rate of geochemical weathering of 0.1 to 0.3 mm year^{-1} as estimated by Yaalon et al. (1984) for a basalt landscape in Israel. They concluded that weathering was contributing around one third to one half of salt input from the rate of rainfall accession at that annual rainfall close to the coast. Trudgill et al. (1989) measured the weathering rates of St. Paul's

Cathedral, London. While the weathering forces of the atmosphere together with the acidic gases dissolved in rainfall will differ from soil, the rates of 0.08 to 0.19 mm year^{-1} are in general agreement with the estimations of Yaalon et al. (1984). Subsoils have higher partial pressures of CO_2, which will enhance weathering. Peck and Hurle (1973) evaluated the rate of chloride release from rock weathering. Assuming a rate of weathering of 0.01 mm year^{-1} they estimated a chloride release rate of around 1 kg ha^{-1} year^{-1}. This compared to the calculated salt fall of 24 to 133 kg ha^{-1} year^{-1} was considered insignificant. Their weathering rate is 10 times smaller than the other estimates, and if the same rate was assumed, the estimated chloride release would be on the order of one half to one tenth of the rainfall input. Velbel (1989) makes the point that weathering rates are easily underestimated by between two and five times, depending on the method of calculation. Streamflow fluxes alone are not appropriate because the biological sink terms are not accounted for appropriately, as stressed by Douglas (1969).

Rhoades et al. (1968) found significant weathering in the soils of the western United States. Shaw et al. (unpublished data) found ongoing weathering in the D horizon of alluvial soils in the Lockyer Valley in southeast Queensland that were derived from basalt. There was release of calcium sufficient to prevent the soils coming to equilibrium with the sodicity of the irrigation water even after 20 years of irrigation. This was attributed to weathering of feldspars and feldspar-clay mineral weathering stages as outlined by McAleese (1958). In both these situations the weathering could be identified by a very high CEC-to-clay ratio, CCR (mEq g^{-1} of clay), in excess of 1.

These situations will result in an overestimate of salt accessions from rainfall. The error will be minimized at shallower depths due to the greater likelihood of salt movement below that depth by rainfall alone, the greater likelihood of active weathering being completed, and the shorter time periods involved in coming to equilibrium with rainfall inputs.

4.3.5.10 Effective Root Zone Depth

From an examination of a large number of soil salt profiles, Shaw and Thorburn (1985a) decided on a soil root zone depth of 0.9 m in an earlier approach to predicting salt leaching from soil properties. The shallower the depth, the less chance there will be of soil weathering making a significant input, but the greater the errors due to water uptake by vegetation. The real problems are in application across a wide range of situations, and the depth chosen should be below the active depth of root water extraction. This will mean that the matrix salt content will reflect the water draining out of the profile that the mass balance equation requires. Two differing situations arise: (1) where a throttle to water movement is shallower than the chosen depth and (2) where water extraction by vegetation occurs below the chosen depth. In the first case this is not a problem, because restricted water entry means limited water available for extraction by plants and thus the salinity of the draining water should reflect the true drainage water composition. The data of Zein el Abedine et al. (1969) show the hydraulic throttle for vertisols in the Gezira to be in the region of 0.2 to 0.7 m in depth. Sukhija et al. (1988) excluded the top 0.9 m, where evapotranspiration effects were dominant. For the second situation, water uptake below the nominated depth will result in errors because the concentration of the drainage water has not attained equilibrium. The errors in this case are small because the proportion of roots at deeper depths is small and the proportional uptake with depth decreases. Error may be large for deep-rooted vegetation. Russell et al. (1967) looked at various soil properties and the growth of Brigalow (*Acacia harpophylla*) on heavy clay soils. They found the greatest weight of roots was in the surface 0.6 m and that relatively insignificant weights of roots would extend below 0.9 m. For trees on permeable soils, deeper depths are obviously applicable, but the shortage of soil data poses constraints for the use of the model. The higher the permeability, the lower the salt content and the smaller the consequent errors.

Thus a depth of 0.9 m is appropriate for many agricultural situations. Shaw and Yule (1978) and Gardner and Coughlan (1982) found rooting depths were generally less than 0.9 m except under highly stressed crops, which had greater rooting depths. Baker and Ahern (1989) looked at published data for rooting depths from a number of Australian studies. Results indicated a rooting depth of between 0.7 and 1.0 m for clay soils, 0.3 and 0.7 m for sodic soils, and up to 1.2 m for better drained soils. There are considerable data for salt profiles for soils in western Australia, for example, Johnston et al. (1980), which show salt peaks over a range of depths down to about 20 m. Sharma (1988) considered 10 m a satisfactory rooting depth for use with trees on a structured laterite. A number of these soils are quite permeable and, as Allison (1988) points out, these profiles tend to reflect quite ancient patterns of wetting events and thus are different from the salt profiles being considered as part of this model. By choosing a shallower depth, the salt profiles are more likely to be in equilibrium with more recent climate in the case of dryland situations, and shorter equilibrium times for irrigated situations. Johnston (1983) suggests least errors will result from sampling just below the root zone, where the change in salt content with depth is zero.

4.3.5.11 Evapotranspiration Rate Is Constant

By combining the steady-state water balance and the steady-state salt balance equations, Shaw (1988b) shows that

$$\frac{c_i}{c_o} = 1 - \frac{ET}{Q_i} \tag{20}$$

Expressing evapotranspiration (ET) and rainfall (Q_i) as fluxes and assuming runoff is negligible, the relationship becomes

$$\frac{c_i}{c_o} = 1 - \frac{\mathrm{ETR}}{K_s} \tag{21}$$

for the root zone where K is saturated hydraulic conductivity. This equation has a similar derivation but from a different basis to that of Bernstein (1967), as described earlier.

The evapotranspiration rate is a most important factor when the soil is wet, because that is the time water movement will be at a maximum. The relative rates of soil water movement downward through the matrix and evapotranspiration are the important criteria. The rate of evapotranspiration is driven by net radiation modified by the rate of removal of water from the evaporating surface. Stewart (1984) considers aerodynamic resistance (roughness) to be more important in removal of water vapor from an evaporating or transpiring surface than was previously thought. This will vary significantly with vegetation density and height. Open forest will have a higher value of evapotranspiration than short grassland. Stewart also states that very little is known about the relative importance of different vegetation types on evapotranspiration in tropical climates. If the temperate pattern persists in the tropics, then different vegetation types will transpire at different rates and the assumption of a constant evapotranspiration rate will not be met. Because there is an increased cloud frequency in wetter areas, there will be a bias toward a lower ETR in higher rainfall areas.

4.4 THE USE OF STREAM SALT MASS BALANCE IN CATCHMENT MANAGEMENT

4.4.1 Introduction

The water carried in rivers and streams is often considered to consist of a proportion of baseflow and a proportion of direct runoff. The base-flow fraction is made up of groundwater that infiltrates into the channel while runoff enters the drainage system during and soon after precipitation periods. Runoff has no residence time in the groundwater system and only a short period of contact with soil or vegetation. However, interactions are extensive enough that the direct runoff has a higher dissolved solids content than the original rain. Baseflow has an even higher dissolved solids content. This means that the solute concentration of river water tends to be inversely related to flow rate. The salt concentration in river water is a product of natural rainfall, erosion, mineralization, and groundwater accessions.

By adapting a catchment-wide approach to stream salinity, it can be rapidly assessed whether the catchment is in equilibrium with salt export or is actively discharging salts via streams. This is an important issue in relation to water-quality control and may also be an indicator of land management effects on surface water quality. Observable changes in surface water quality may result from long-term changes in rainfall or runoff cycles, or from changes in land use or in water utilization.

4.4.2 Sampling Intensity of Streams

Continuous information on dissolved salts via records of flow and concentration is available only in rare cases. Manual sampling programs provide the basis for many studies of salt transport and it is important to recognize the limitations associated with infrequent sampling frequency. For an accurate assessment of salt export, it would be essential to take daily water-quality samples. To determine adequately the composition of a flowing stream, each sample, or set of samples taken simultaneously, must be representative of the entire flow at the sampling point at that instant. Furthermore, the sampling process must be repeated frequently enough to define changes with time that may occur in the water passing the sampling point. A single grab sample ought to be considered only to represent the instantaneous discharge at the time of sampling.

However, because such intensive sampling is rarely undertaken, it is more appropriate to model the relationship between water flow and salt concentration in such a way that total salt loads can be calculated from daily flow information. This will be discussed further in later sections. The determination of sources of salt input, and processes involved in salt export from catchments, can be made using simplified salt mass balance concepts.

4.4.3 Salt Input from Rainfall

Rainfall is the major contributing factor to streamflow in many inland Australian streams, which tend to be ephemeral in nature. The concentration of salts in rainfall varies isometrically from the coast, but for inland areas can be considered relatively constant on a regional basis. Ayers (1982) related the chloride concentration in the soil profile to the accession of chloride in rainfall. Other work has also highlighted the gradient in salt concentration of rainfall from the coast (Blackburn and McLeod, 1983). However, the gradient of salt concentration declines dramatically, such that at distances greater than 200 km from the coast rainfall can be considered essentially uniform in salt concentration. When the distance to the coast exceeds 100 km, the oceanic influence is gradually superseded by continental inputs of salt to the atmosphere from the dissolution of soil dust and forest fire material (Meybeck, 1983). Major differences in salt concentration of rainfall are related to storm intensity but yearly averages show a closer relationship to distance from the coast and prevailing winds. There are other sources of dissolved salts apart from the natural cycling, especially if an area has been intensively developed. Rural areas in Australia are generally sparsely populated so that artificial inputs into streams are minimal. Therefore, if we compare the total salt input from rainfall over time with total output via streamflow, an assessment of catchment salt balance can be made.

Chloride is a highly soluble element that is commonly not present in rocks, although a little can be released during rock weathering. Thus the stream content of chloride will commonly be simply related to the average rainfall over the drainage basin and its associated runoff. Chloride is chosen as the reference element for studying atmospheric influence on river chemistry as very little is released by rock weathering. In most cases, it is chloride salts that are of most concern in water-quality decisions, especially with agricultural waters. Therefore it is appropriate to use chloride in a salt mass balance approach to catchment management, as this element will directly reflect changes in the water balance due to changes in land use or management.

There are other constituents of rainfall apart from chlorides, so that it was considered approprate to use the electrical conductivity (EC) of rainfall as a measure of dissolved salt content. The use of electrical conductivity measurements is also commonly used for determining the concentration of salts in rivers and streams; hence for consistency of analysis, EC is used in the calculation of salt mass balance.

4.4.4 Assumptions in Use of Steady-State Salt Balance Approach

The assumptions involved in the use of steady-state salt balance models were discussed in previous sections of this chapter. This approach has been extrapolated to a larger scale to provide a sensitivity-type analysis of catchments, which can be used to define those catchments that are more sensitive to land use change. These catchments should then be the subject of further research and refined modeling techniques to define solute movement pathways more thoroughly.

Steady-state salt balance assumes that there is uniform application of water and salts over the entire catchment; that salt is conserved in the system; and any land management changes create a change that has moved well toward a new equilibrium. Salt input into the system is considered constant from rainfall, and all the salt is available for movement, i.e., it is considered to be nonreactive. Although groundwater inflow into streams is commonly small, the movement of groundwater must be restricted to within the catchment of interest. This assumption is unlikely to be fully met as groundwater systems are commonly regional in extent. However, if the movement of solutes in groundwater is several orders of magnitude slower than in surface water, as is not uncommon, it will have little influence for analysis over shorter time periods, apart from acting as a potential buffer for salts.

4.4.5 Catchment Mass Salt Balance

Broader assumptions used when calculating the salt load in streams from discharge information include the following: soil and rock mineralization contribute minimally; groundwater solute concentration and inflow is defined by low flow discharge and quality measurements; and rainfall solute concentration is constant across the catchment with uniform salt deposition. These assumptions simplify the model of salt movement in the catchment into a steady-state mass balance where input is via rainfall, and output is via streamflow.

This approach provides us with a "leaky" bucket-type analogy of mass salt balance for the catchment where there is a measured rate of water and salt movement out of the bottom of the bucket. If the bucket is full, the amount of salt entering the system must be equivalent to the amount leaving the system, via increased salt loads in streams. If the bucket is empty, the system is capable of storing more salts and there will be little change in the quality of streams, until the bucket is eventually filled. It may be that a new equilibrium is reached where the bucket is never filled, so that land use changes will have limited impact on stream water quality. However, in many cases the important consideration is the time taken to fill the bucket after a land use change, which can lead to the eventual decline in soil and water quality.

The approach outlined in this section is not intended to make predictions; it is simply a method for determining, from historical water flow and limited water-quality data, the status quo of a catchment in terms of a salt mass balance. For those interested in the effect of land development on water quality, this approach will highlight those catchments where salinity is more likely to respond to land use change, and which will therefore require more detailed study.

4.4.6 Salt Export in Streams

The difference between salt input (predominantly chloride) via rainfall, and export via streams, can be attributed to a number of factors. If salt export is significantly greater than rainfall input, it would indicate that landscape processes were actively producing soluble salts, or the groundwater system was discharging saline water into streams. Especially in inland rivers and streams of Australia the baseflow component due to groundwater discharge is only a very small percentage of the total annual flow.

If the export of salts is significantly less than rainfall accessions, it would indicate a catchment landscape capable of absorbing soluble salts, or alternatively a landscape capable of moving salts into a groundwater system not linked to surface waters, or a groundwater system so large that the dilution effect is great.

Following this scenario of examining a rough salt balance for catchments, it is possible to gain an indication of the sensitivity of a catchment to land use changes. If a catchment that has been cleared for 30 years or more is in equilibrium, we could assume a relatively constant salt load over time. If a catchment is in equilibrium prior to clearing, then it may be sensitive to future land development. However, if a catchment is still "accumulating" salt it may be able to absorb the impact of land development without much influence on surface water quality.

4.4.7 Water Quality vs. Streamflow

Surface water quality is related to the relevant proportions of dissolved salts, suspended materials, and industrial and agricultural chemical residues in rivers and streams. Previous work in Australia (Schofield, 1988) has found the level of salts in surface water is a measure of the total salts dissolved from the soil strata. During the 1970s an interest developed in determining the relationship between dissolved salts and discharge rates in streams (Hall, 1970, 1971). In general, studies involved regression analysis or best line fit of the concentration of dissolved salts and stream discharge, and such analysis was considered adequate for water-quality predictions. This approach does little to interpret the relationship between flow and quality, and is simply a method for grouping data.

Work in the United States (Hem, 1970) indicated that the range between high and low extremes of dissolved salt concentrations for a daily composited stream sample was rarely as wide as that between high and low flow rates. This indicates the problems of compositing samples and averaging results, which ignores daily fluctuations in the flow-quality relationship for many streams. Usually the changes in dissolved solids concentration are related both to the rate of water discharge and to the rate of change of discharge, but this relationship is complicated by other factors such as the buffering capacity of drainage systems. Hem (1970) suggested that for some streams, past records can be directly utilized as a means of estimating water quality from discharge, while more intensive sampling is required for some streams.

Many authors recognized that meaningful estimates of total dissolved load cannot be obtained from the product of mean annual water discharge and mean concentration (Davis and Zobrist, 1978; Hall, 1970; Hem, 1970; Walling and Webb, 1983). Significant overestimation of load will result in nearly all cases using this method of estimation. Walling and Webb (1983) have documented information on total dissolved load from 490 rivers throughout the world to review the general characteristics of dissolved salt transport.

4.4.8 Predicting Salt Loads in Streams

Different approaches to evaluating surface water data are outlined by Davis and Keller (1983) and can be summarized as follows.

Time course — A time course is simply a plot of salt concentration as a function of time, giving an indication of possible patterns in the fluctuation of this parameter. An asymmetrical sinusoidal behavior is often observed in salt concentration, but with little correlation to the corresponding changes in water flow.

Average values — Averages provide a simple and rapid overview of large amounts of data, but do not provide information on fluctuations.

Cumulative distribution curves — Cumulative distribution curves provide information on frequency of occurrence of values over the entire range observed, but do not set changes in the magnitude and distribution of concentrations into the context of corresponding flow variations.

Correlation analysis — The correlation analysis of salt concentration and water flow allows reduction of data and gives an indication of the interdependence, although not necessarily the direction of causation, between two parameters. However, if the relationship between parameters is affected by seasonal factors, there may be no detectable correlation over the period of an entire year.

Factor analysis — Factor analysis involves grouping of those parameters that have related behavior and can indicate in one analysis which factors, in addition to flow, influence concentration (e.g., date of sampling and geological source).

Waterflow (Q) vs. concentration (c) diagrams — Q vs. c diagrams enable the influence of flow on concentration to be directly visualized. Using this relationship for regression analysis makes the assumption that the independent variable (Q) exerts the main influence on the dependent variable (c).

Flow-concentration-time (Q-c-t) diagrams — The three-dimensional Q-c-t relationship allows the concentration to be examined as a function of flow during the course of time. Thus different concentrations, related to the same discharge level occurring at different times of the year, can be readily identified.

Mathematical models for defining salt loads in streams have been developed by Davis and Zobrist (1978) in Europe, and by Hall (1970, 1971), Johnson et al. (1969), Johnson and Swank (1973), and Johnson (1979) in the United States. In some cases biological models have been adapted to suit the description of water quality vs. discharge phenomena (e.g., Landsberg, 1977).

A brief overview of mathematical approaches used to describe salt concentration in streams is given as follows:

Log-log relationships — Most studies have used log-log transformations with least-squares regression to develop a simple function of the form

$$c = aQ^b \tag{22}$$

where c is solute concentration, Q is discharge, a is the regression constant, and b is the regression exponent. It has been found that the exponent b of the rating relationship has a negative value, which indicates a decrease in solute concentration with increasing discharge from a sampling station (Walling and Webb, 1986).

Biological models — An adoption of simple biological models has been made to describe flow/concentration relationships (Landsberg, 1977), an example being

$$Y = Y_{\max}\left[1 + \left(X/X_{0.5}\right)^m\right] \tag{23}$$

where Y is the dissolved salt concentration (T.D.S), $Y_{\max}$ is the maximum TDS in the measurement period, X is the flow, $X_{0.5}$ is the flow at 50% TDS and m is a constant.

Mixing models — A combination of simple mass balances and mixing models provides dissolved solids-discharge relationships that are not completely empirical, but their use involves assumptions that place limitations on their applicability (Hall, 1970, 1971). An example is

$$c = \frac{S}{1 + BQ^{1/m}} \tag{24}$$

where S is the dissolved solids concentration, Q is discharge concentration, and B and m are constants.

Models used to describe stream salinity in Australia have been developed to predict changes resulting from tree clearing or other land use modifications. Western Australia has been particularly active in this sphere with predictive models developed by Schofield (1988), Loh and Stokes (1981), and Mauger (1986). Predictive models simulate generation of streamflow and stream salinity from data on rainfall, evaporation, and physical characteristics of a catchment. These models all require input of a number of physical catchment parameters that require a fair degree of experimental effort to collect. For this reason a mathematical model that requires limited data collection is more suitable for a reconnaissance-style approach to catchment salt loads. The most suitable models found to fit Australian, and more specifically Queensland, inland rivers, were mathematical relationships developed on a mass balance concept. A review of literature from the United States (Walling and Webb, 1983) indicated that, in general, a plot of water discharge vs. dissolved solids concentration (as measured by conductance) is commonly hyperbolic, with concentration reaching a limiting value at low discharge near the concentration of the inflow source (baseflow) and a limiting value at high discharge fixed by the concentration of salt in rainwater plus the solutes added by the effect of rapid reactions of the runoff with exposed rock and soil minerals. Hall (1970) developed a series of dissolved solids-discharge relationships derived on the basis of simple mass balances and mixing models. One of these equations was adapted to data available on several

Queensland streams to define more appropriately the salt load in these rivers. The chosen equation was of the following form:

$$c = \frac{S - c_i}{1 + BQ^{1/m}} + c_i \tag{25}$$

where c is the concentration of dissolved solids in the stream, S is the base-flow concentration (taken as TDI at lowest flow), c_i is the inflow concentration (taken as rainfall), Q is the discharge, and B and m are constants.

4.4.9 Catchment Salt Balance in Queensland

Daily discharge information was available for many inland rivers in southern Queensland. These rivers fed directly into the Murray-Darling system, which is the catchment with the largest irrigated area in Australia. There is considerable interest in the level of salt being exported into the Murray-Darling system and the possible consequences of changes in land management in the upper catchment. As part of this exercise an equation was adapted from the work of Hall (1970) to describe the relationship between stream flow and stream water quality [Equation (25)]. Limited water quality information was used to calibrate the equation for an individual river. As is common in most countries, daily water flow information is available, but only limited sampling for water quality is undertaken.

Figure 4.10 displays water-quality vs. flow data for McIntyre Brook, an ephemeral stream in Queensland. Fitted to these data is a simple arithmetic average (A), a log fit (B), and the modified Hall equation (C). Future refinement of Equation (25) may be necessary, but streams could be rated according to this equation, using historic data. All these approaches have various limitations. Arithmetic averages tend to overestimate the salt loads due to the relationship between discharge rate and salt concentration. The use of log relationships will underestimate the concentration of dissolved salts at high stream discharge levels. Equation (25) has limitations due to a lack of data in the high-discharge range, which makes it difficult to accurately assign end points for the equation. Information available for most Australian rivers gives an abundance of data for medium- to low-discharge events, but very few data for high-discharge events.

The relationship between water quality and discharge may be further clarified by separating data from advance and recessionary flows. A preliminary analysis of data for Queensland rivers indicates that two different hyperbolic curves of similar form to that described above could be fitted to advance phases and recession phases of flow. The concentration of dissolved solids in advance phases is higher for a given discharge than for recession phases of flows. This would indicate a flushing of the drainage system after significant rainfall events, followed by a return to rainfall-driven equilibrium conditions. Developments of this type may be used for more advanced pollution studies. However, fitting of a single equation such as Equation (25) to historic discharge data will provide a significant advance on current practice of using monthly or yearly averages to rate rivers and streams in terms of water quality.

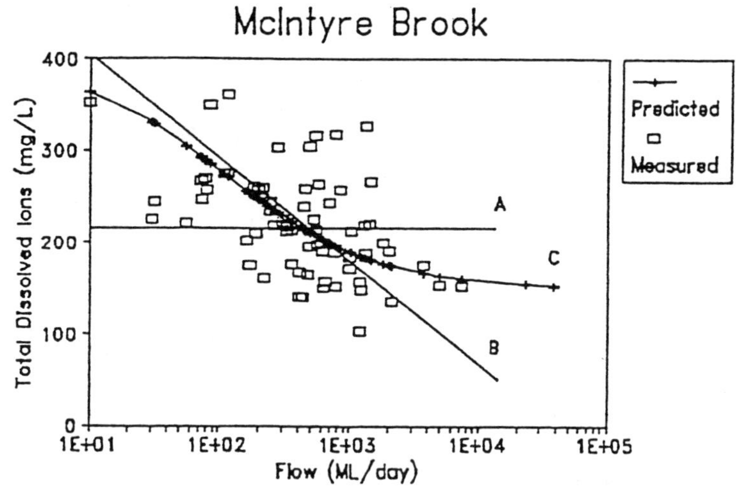

FIGURE 4.10. The relationship between stream flow and water quality for McIntyre Brook.

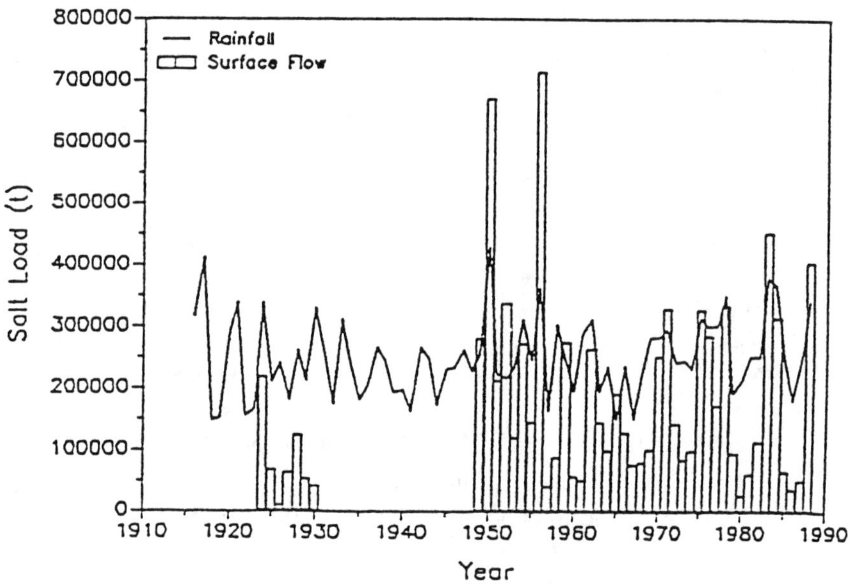

FIGURE 4.11. Catchment mass salt balance for the McIntyre River in Queensland, Australia. (—), Rainfall; (□), surface flow.

Figure 4.11 displays the salt balance for the same stream, McIntyre Brook in Queensland, as is the subject of Figure 4.10. In this catchment the amount of salt deposited in rainfall is in rough equilibrium with stream discharge down the river. This catchment fits in with the full bucket analogy, where any changes to the system may result in excess salts being discharged into streams. This predicted scenario is supported by data on several smaller tributaries within the catchment that have become increasingly saline after extensive tree clearing. The removal of trees has led to an increase in stream discharge and salt load due to a combination of increased baseflow contributions from an elevated water table and an increase in ephemeral surface discharge. The increased export of water and salts from a catchment after tree clearing is due to the difference in water use between native tree species and the crops and/or pastures that are introduced with land development for agriculture (Ruprecht and Schofield, 1989, 1990). This situation is quite different from historically saline streams, where a mass balance would indicate an equilibrium with rainfall over long periods. Changes in stream salinity would be represented by a change in the input-discharge relationship, hence there would be a convergence or divergence of the rainfall and streamflow concentration lines.

4.5 LIST OF SYMBOLS

a	Regression coefficient
b	Regression coefficient
B	Experimental constant in mixing model relationships
c	Solute concentration, $M\ L^{-3}$
c_i	Solute concentration of infiltrating water, $M\ L^{-3}$
c_n	Mass of solute per unit area stored in the soil profile, $M\ L^{-2}$
c_o	The concentration of electrolyte in the water draining below the root zone, commonly represented by electrical conductivity, EC
c_0	Initial solute concentration, $M\ L^{-3}$
$\bar{c}$	Average solute concentration of the soil water to a depth, $M\ L^{-3}$
$\bar{c}_f$	Steady-state value of $\bar{c}$
$\bar{c}_0$	Value of $\bar{c}$ at time $t = 0$
C_w	The weighted salt concentration over the nominated root zone depth expressed as EC_{se}
C_w^*	The appropriate salt tolerance rating for a particular crop expressed as EC_{se}
D_r	The daily drainage rate over the time period considered, LT^{-1}
D_d	The depth of water draining below the root zone, equivalent to Q_0, L
D_e	Depth of water evapotranspired, L
D_i	The depth of applied water, equivalent to Q_i, L
D	Apparent dispersion coefficient, $L^2\ T^{-1}$
D_0	Molecular diffusion coefficient, $L^2\ T^{-1}$
E	Evapotranspiration rate, $L\ T^{-1}$
EC	Electrical conductivity of a solution, usually in units of dS m^{-1}
EC_{se}	Electrical conductivity of a soil saturation extract, in units of dS m^{-1}
ESP	Exchangeable sodium percentage at the bottom of the root zone, M
ET	Evapotranspiration, L
ETR	Evapotranspiration rate, $L\ T^{-1}$

I	Infiltration rate, $L\ T^{-1}$
k	A coefficient that linearizes the influence of ESP
K	Hydraulic conductivity, $L\ T^{-1}$
K_s	Saturated hydraulic conductivity of soil, $L\ T^{-1}$
L	Leaching flux past a depth, $L\ T^{-1}$
LF	Leaching fraction
LR	Leaching requirement
n	Time element "counter"
m	Empirical constant in mixing model relationships
Q	Stream discharge, $L^3\ T^{-1}$
Q_F	The "leaching constant" that accounts for the soil properties determining the release of salts, L
Q_i	The quantity of water entering the soil, L
Q_o	The quantity of water draining below the root zone, L
q	Darcian water flux, $L\ T^{-1}$
R	Depth of rooting, L
R_m	Depth of median annual rainfall, L
S	Baseflow solute concentration, $M\ L^{-3}$
s_i	The total initial salt content in the soil, $M\ L^{-2}$
s_o	the salinity of the soil after leaching, $M\ L^{-2}$
t	Time, T
t_i	The time period of infiltration over the irrigation cycle, T
u	Amount of solute taken up by a crop, $M\ L^{-2}$
v	Pore water velocity = q/θ, $L\ T^{-1}$
v_m	Mean pore water velocity = $\Delta\alpha/\Delta t$, $L\ T^{-1}$
$W_{n,z}$	Equivalent ponded depth of water that passed below a given depth, L
X	Independent variable in biological models
Y	Dependent variable in biological models
Y_{max}	Maximum value for dependent variable Y
z	Soil depth, L
α	Depth beneath the soil surface of a solute concentration peak, L
β	Depth of penetration of the step decrease in solute input = $\alpha - \tau$, L
γ	Depth over which the soil water deficit must be replenished for peak movement to occur, L
ε	Coefficient of solute dispersivity in soil, L
θ	Volumetric water content, $L^3\ L^{-3}$
θ_{fc}	Volumetric water content at "field capacity," $L^3\ L^{-3}$
θ_ℓ	Volumetric water content at which leaching occurs, $L^3\ L^{-3}$
κ	Proportion of evapotranspiration occurring before infiltration
λ	Dimensionless parameter = $c_z/\bar{c}$
τ	Depth of the solute pulse = $I_n\Delta t/\theta_{fc}$, L
Φ	Dimensionless coefficient to account for crop solute uptake = $(c_n - u)/c_n$
ψ	Soil matric potential, L

4.6 REFERENCES

Addiscott, T. M. 1977. A simple computer model for leaching in structured soils. *Soil Sci.* 28:554–563.

Addiscott, T. M. and R. J. Wagenet. 1985. Concepts of solute leaching in soils: a review of modeling approaches. *J. Soil Sci.* 36:411–424.

Addiscott, T. M., P. J. Heys, and A. P. Whitmore. 1986. Application of simple leaching models in heterogeneous soils. *Geoderma* 38:185–194.

Allison, G. B. 1988. A review of some of the physical, chemical and isotopic techniques available for estimating groundwater recharge. In: Estimation of Natural Groundwater Recharge. I. Simmers (Ed.). NATO ASI Ser. C, *Math. Phys. Sci.* 222:49–72.

Allison, G. B. and M. W. Hughes. 1978. The use of environmental chloride and tritium to estimate total recharge to an unconfined aquifer. *Aust. J. Soil Sci.* 16:181–195.

Allison, G. B. and M. W. Hughes. 1983. The use of natural tracers as indicators of soil-water movement in a temperate semi-arid region. *J. Hydrol.* 60:157–173.

Allison, G. B., W. J. Stone, and M. W. Hughes. 1985. Recharge in karst and dune elements of a semi-arid landscape as indicated by natural isotopes and chloride. *J. Hydrol.* 76:1–25.

Appelt, H., K. Holtzclaw, and P. F. Pratt. 1975. Effect of anion exclusion on the movement of chloride through soils. *Soil Sci. Soc. Am. Proc.* 39:264–267.

Ayers, G. P. 1982. The chemical composition of precipitation: a southern hemisphere perspective. In: Atmospheric Chemistry. E. D. Goldberg (Ed.). Springer-Verlag, New York, pp. 41–56.

Baker, D. E. and C. A. Ahern. 1989. Estimates of effective rooting depth for predicting available water capacity of burdekin soils, Queensland. *Aust. J. Soil Res.* 27:438–454.

Barraclough, D., M. J. Hyden, and G. P. Davies. 1983. Fate of fertiliser nitrogen applied to grassland. 1. Field leaching results. *J. Soil Sci.* 34:483–497.

Barry, D. A., C. W. Rose, P. G. Saffigna, and J. Y. Parlange. 1985. Interpretation of leaching under multiple fertiliser applications. *J. Soil Sci.* 36:9–20.

Beese, F. and P. J. Wierenga. 1980. Solute transport through soil with adsorption and root water uptake computed with a transient and a constant flux model. *Soil Sci.* 129:245–252.

Bernstein, L. 1967. Quantitative assessment of irrigation water quality. In: Water Quality Criteria. Special publication No. 416. American Society for Testing and Materials, pp. 51–64.

Bernstein, L. and L. E. Francois. 1973. Leaching requirement studies: sensitivity of alfalfa to irrigation and drainage waters. *Soil Sci. Am. Proc.* 37:931–943.

Biggar, J. W. and D. R. Nielsen. 1976. Spatial variability of the leaching characteristics of a field soil. *Water Resour. Res.* 12:78–84.

Blackburn, G. and S. McLeod. 1983. Salinity of atmospheric precipitation in the Murray-Darling drainage division, Australia. *Aust. J. Soil Res.* 21:411–434.

Blackmore, A. V. 1976. Salt sieving within soil aggregates. *Aust. J. Soil Sci.* 14:149–158

Bolt, G. H. 1979. Movement of solutes in soils: principles of adsorption/exchange chromatography. In: Soil Chemistry. B. Physico-chemical Models, Ch. 9. G. H. Bolt (Ed.) Elsevier, Amsterdam, pp. 285–348.

Bond, W. J. and D. E. Smiles. 1988. Predicting the average movement of reactive solutes in soils. *Soil Use Management* 4:115–120.

Bouma, J., L. W. Dekker, and C. J. Muilwijk. 1981. A field method for measuring short-circuiting in clay soils. *J. Hydrol.* 52:347–354.

Boumans, J. H. 1963. Some principles governing the drainage and irrigation of saline soils. In: Reclamation of Salt Affected Soils in Iraq; Soil Hydrological and Agricultural Studies. Publication No. 11. P. J. Dieleman (Ed.). International Institute for Land Reclamation and Improvement, Wageningen.

Bresler, E. 1973. Simultaneous transport of solute and water under transient unsaturated flow conditions. *Water Resour. Res.* 9:975–986.

Bresler, E. and G. Dagan. 1979. Solute dispersion in unsaturated heterogeneous soil at field scale. II. Applications. *Soil Sci. Soc. Am. J.* 43:467–472.

Bresler, E. and G. Dagan. 1981. Convective and pore scale dispersive solute transport in unsaturated heterogeneous fields. *Water Resour. Res.* 17:1683–1693.

Bresler, E. and R. J. Hanks. 1969. Numerical method for estimating simultaneous flow of water and salt in unsaturated soils. *Soil Sci. Soc. Am. Proc.* 33:827–832.

Burch, G. J., R. K. Bath, I. D. Moore, and E. M. O'Loughlin. 1987. Comparative hydrological behaviour of forested and cleared catchments in south eastern Australia. *J. Hydrol.* 90:19–42.

Burns, I. G. 1975. An equation to predict the leaching of surface-applied nitrate. *J. Agric. Sci.* 85:443–54.

Cameron, D. R. and A. Klute. 1977. Convective-dispersive solute transport with a combined equilibrium and kinetic adsorption model. *Water Resour. Res.* 13:183–188.

Cameron, K. C. and A. Wild. 1982. Prediction of solute leaching under field conditions: an appraisal of three methods. *J. Soil Sci.* 33:659–669.

Campbell, G. S. 1974. A simple method for determining unsaturated conductivity from moisture retention data. *Soil Sci.* 117:311–314

Cassel, D. K., M. Th. Van Genuchten, and P. J. Wierenga. 1975. Predicting anion movement in disturbed and undisturbed soils. *Soil Sci. Soc. Am. J.* 39:1015–1019.

Dagan, G. and E. Bresler. 1979. Solute dispersion in unsaturated heterogeneous soil at field scale. I. Theory. *Soil Sci. Soc. Am. J.* 43:461–466.

Davis, J. S. and H. M. Keller. 1983. Dissolved loads in streams and rivers — discharge and seasonally related variations. In: Dissolved Loads of Rivers and Surface Water Quantity/Quality Relationships, Proc. of the Hamburg Symposium, August. International Association of Hydrological Sciences, Publication No. 141, pp. 79–89.

Davis, J. S. and J. Zobrist. 1978. The interrelationships among chemical parameters in rivers — analysing the effect of natural and anthropogenic sources. *Prog. Water Technol.* 10(5–6):65–78.

De Smedt, F. and P. J. Wierenga. 1978a. Solute transport through soil with nonuniform water content. *Soil Sci. Soc. Am. J.* 42:7–10.

De Smedt, F. and P. J. Wierenga. 1978b. Approximate analytical solution for solute flow infiltration and redistribution. *Soil Sci. Soc. Am. J.* 42:407–412.

Dettinger, M. D. 1989. Reconnaissance estimates of natural recharge to desert basins in Nevada, USA, by using chloride-balance calculations. *J. Hydrol.* 106:55–78

Devitt, D., W. M. Jarrell, W. A. Jury, O. R. Lunt, and L. H. Stolzy. 1984. Wheat response to sodium uptake under zonal saline-sodic conditions. *Soil Sci. Soc. Am. J.* 48:86–92.

Douglas, I. 1969. The efficiency of humid tropical denudation systems. *Inst. Br. Geogr. Trans.* 46:1–16.

Dowling A. J., P. J. Thorburn, P. J. Ross, and P. J. Elliot. 1991. Estimation of infiltration and deep drainage in a furrow irrigated duplex soil. *Aust. J. Soil Res.* 29:363–375.

Dudley, L. M., R. J. Wagenet, and J. J. Jurinak. 1982. Description of soil chemistry during transient solute transport. *Water Resour. Res.* 17:1498–1504.

Dyson, J. S. and R. E. White. 1987. A comparison of the convection-dispersion equation and transfer function model for predicting chloride leaching through an undisturbed, structured clay soil. *J. Soil Sci.* 38:157–172.

Edmunds, W. M., W. G. Darling, and D. G. Kinniburgh. 1988. Solute profile techniques for recharge estimation in semi-arid and arid terrain. In: Estimation of Natural Groundwater Recharge. *NATO ASI Ser. C, Math. Phys. Sci.* 222:139–157.

Elrick, D. E., B. E. Clothier, and J. E. Smith. 1987. Solute transport during absorption and infiltration: a comparison of analytical approximations. *Soil Sci. Soc. Am. J.* 51:282–287.

Eriksson, E. and V. Khunakasem. 1969. Chloride concentration in groundwater, recharge rate and rate of deposition of chloride in the Israel coastal plain. *J. Hydrol.* 7:178–197.

Frenkel, H., J. O. Goertzen, and J. D. Rhoades. 1978. Effects of clay type and content, exchangeable sodium percentage, and electrolyte concentration on clay dispersion and soil hydraulic conductivity. *Soil Sci. Soc. Am. J.* 42:32–39.

Gardner, E. A. and K. J. Coughlan. 1982. Physical factors determining soil suitability for irrigated crop production in the Burdekin-Elliot River area. Agricultural Chemistry Branch Technical Report No. 20. Queensland Department of Primary Industries, Queensland, Australia.

Gardner, W. R. 1965. Movement of nitrogen in soil. In: Soil Nitrogen. Agronomy No. 10. W. V. Bartholomew and F. E. Black (Eds.). American Society of Agronomists, Madison, WI. pp. 550–572.

Gardner, W.R. 1967. Water uptake and salt distribution patterns in saline soils. Proc. Symp. on Isotope and Radiation Techniques in Soil Physics and Irrigation Studies. Int. Atomic Energy Agency, Istanbul, June 1967, pp. 335–341.

Gee, G. W. and D. Hillel. 1988. Groundwater recharge in arid regions: review and critique of estimation methods. *Hydrol. Proc.* 2:255–266.

Gupta, S. P. and R. A. Greenkorn. 1974. Determination of dispersion and nonlinear adsorption parameters for flow in porous media. *Water Resour. Res.* 10:839–849.

Gvirtzman, H., N. Paldor, M. Margaritz, and Y. Bachmat. 1988. Mass exchange between mobile freshwater and immobile saline water in the unsaturated zone. *Water Resour. Res.* 24:1638–1644.

Hall, F. R. 1970. Dissolved solids-discharge relationships. 1. Mixing models. *Water Resour. Res.* 6(3):845–851.

Hall, F. R. 1971. Dissolved Solids-Discharge Relationships. 2. Applications to Field Data. Institute of Natural and Environmental Resources, University of New Hampshire, Durham, NH.

Hamid, A. and B. P. Warkentin. 1973. Effect of aggregate size on leaching of salts from soil aggregates. *Pak. J. Sci. Ind. Res.* 16:54–55.

Hantschel, R., M. Kaupenjohann, R. Horn, J. Gradl, and W. Zech. 1988. Ecologically important differences between equilibrium and percolation soil extracts, Bavaria. *Geoderma* 43:213–227.

Hem, J. D. 1970. Study and Interpretation of the Chemical Characteristics of Natural Water, 2nd ed. U.S. Government Printing Office, Washington, D.C.

Hillel, D. and R. S. Baker. 1988. A descriptive theory of fingering during infiltration into layered soils. *Soil Sci.* 146:51–56.

Hoffman, G. J. and M. Th. van Genuchten. 1983. Soil properties and efficient water use: water management for salinity control. In: Limitations to Efficient Water Use in Crop Production. H. M. Taylor, W. R. Jordan, and T. R. Sinclair (Eds.). American Society of Agronomy, Madison, WI, pp. 73–85.

Hutton, J. T. and T. I. Leslie. 1958. Accession of non-nitrogenous ions dissolved in rainwater to soils in Victoria. *Aust. J. Agric. Res.* 9:492–507.

Isbell, R. F., R. Reeve, and J. T. Hutton. 1983. Salt and sodicity. In: Soils: An Australian Viewpoint. Division of Soils, Commonwealth Scientific and Industrial Research Organisation (CSIRO Melbourne/Academic Press, London), pp. 107–117.

Jacobsen, T. and R. M. Adams. 1958. Salt and silt in ancient Mesopotamian agriculture. *Science* 128:1251–1258

Jarvis, N. J., P. B. Leeds-Harrison, and J. M. Dosser. 1987. The use of tension infiltrometers to assess routes and rates of infiltration in a clay soil. *J. Soil Sci.* 38:633–640.

Jensen, J. R. 1983. Chloride dispersion in packed columns during saturated flow. *J. Soil Sci.* 34:249–262.

Johnson, A. H. 1979. Estimating solute transport in streams from grab samples. *Water Resour. Res.* 6(3):591–601.

Johnson, N. M., G. E. Likens, F. H. Borman, D. W. Fisher, and R. S. Pierce. 1969. A working model for the variation in stream water chemistry at the Hubbard Brook experimental forest, New Hampshire. *Water Resour. Res.* 5(6):1353–1363.

Johnson, P. L. and W. T. Swank. 1973. Studies of cation budgets in the southern Appalachians on four experimental watersheds with contrasting vegetation. *Ecology* 54(1):70–80.

Johnston, C. D. 1983. Estimation of groundwater recharge from the distribution of chloride in deeply weathered profiles from south-west Western Australia, International Conference on Groundwater and Man, Sydney, Australia, pp. 143–152.

Johnston, C. D. 1987. Distribution of environmental chloride in relation to subsurface hydrology. *J. Hydrol.* 94:67–88.

Johnston, C. D., McArthur, and A. J. Peck. 1980. Distribution of soluble salts in soils of the Manjimup Woodchip Licence Area, Western Australia. Land Resources Management Technical Paper No. 5. Commonwealth Scientific and Industrial Research Organisation Australia.

Johnston, C. D., D. H. Hurle, D. R. Hudson, and M. I. Height. 1983. Water movement through preferred paths in lateritic profiles of the Darling Plateau, Western Australia. Groundwater Research Technical Paper No. 1. Commonwealth Scientific and Industrial Research Organization Australia.

Jury, W. A. 1982. Simulation of solute transport function model. *Water Resour. Res.* 18:363–368.

Jury, W. A. 1983. Chemical transport modeling: current approaches and unresolved problems. In: Chemical Mobility and Reactivity in Soil Systems. pp. 49–64.

Jury, W. A. 1984. Field scale water and solute transport through unsaturated soils. In: Soil Salinity Under Irrigation. *Ecol. Stud.* 51:115–125.

Jury, W. A. 1988. Solute transport and dispersion. In: Flow and Transport in the Natural Environment, Proc. Inter. Symp. Canberra, Australia, September 1987. W. L. Steffen and O. T. Denmead (Eds.). Springer-Verlag, Berlin, pp. 1–16

Jury, W. A. and P. F. Pratt. 1980. Estimation of the salt burden of irrigation waters. *J. Environ. Qual.* 9:141–146.

Jury, W. A. and G. Sposito. 1985. Field calibration and validation of solute transport models for the unsaturated zone. *Soil Sci. Soc. Am. J.* 49:1331–1341.

Jury, W. A., W. R. Gardner, P. G. Saffigna, and C. B. Tanner. 1976. Model for predicting simultaneous movement of nitrate and water through a loamy sand. *Soil Sci.* 122:36–43.

Jury, W. A., L. H. Stolzy, and P. Shouse. 1982. A field test of the transfer function model for predicting solute transport. *Water Resour. Res.* 18:369–375.

Jury, W. A., G. Sposito, and R. E. White. 1986. A transfer function model of solute transport through soil. 1. Fundamental concepts. *Water Resour. Res.* 22:243–247.

Khanif, Y. M., O. Van Cleemput, and L. Baert. 1984. Evaluation of the burns model for nitrate movement in wet sandy soils. *J. Soil Sci.* 35:511–518.

Kirda, C., D. R. Nielsen, and J. W. Biggar. 1973. Simultaneous transport of chloride and water during infiltration. *Soil Sci. Soc. Am. Proc.* 37:339–345.

Kirda, C., J. L. Starr, C. Misra, J. W. W. Biggar, and D. R. Nielsen. 1974. Nitrification and denitrification during miscible displacement in unsaturated soil. *Soil Sci. Soc. Am. Proc.* 38:772–776.

Kirkby, M. J. 1985. A basis for soil profile modeling in a geomorphic context. *J. Soil Sci.* 36:97–121.

Kneale, W. R. and R. E. White. 1984. The movement of water through cores of a dry (cracked) clay-loam grassland topsoil. *J. Hydrol.* 67:361–365.

Landsberg, J. J. 1977. Some useful equations for biological studies. *Exp. Agric.* 13:273–286.

Lawrence, P. A. and D. P. Sinclair. 1989. Catchment hydrology and soil loss. In: Changes in Hydrology, Soil Fertility and Productivity of Brigalow Catchments following Clearing. Report RQR89002. P. A. Lawrence and P. J. Thorburn (Eds.). Queensland Department of Primary Industries, Queensland, Australia, pp. 2.1–2.98.

Loh, I. and R. A. Stokes. 1981. Predicting stream salinity changes in southwestern Australia. *Agric. Water Manage.* 4:227–254.

Lyle, C. W., A. H. Mehanni, and A. P. Repsys. 1986. Leaching rates under a perennial pasture irrigated with saline water. *Irrig. Sci.* 7:277–286.

Madan Kumar, N. and M. C. Oswal. 1984. Test on one-dimensional models for predicting salt dynamics in soils. *Soil Sci.* 137:408–414.

Matthias, A. D., H. M. Hassan, Yu-Qi Hu, J. E. Watson, and A. W. Warrick. 1986. Evaporotranspiration estimates derived from subsoil salinity data. *J. Hydrol.* 85:209–223.

Mauger, G. W. 1986. Darling Range Catchment Model. Conceptual Model, Vol. 1. Report No. WP9. Water Resources Planning Branch, Water Authority of Western Australia, Perth.

McAleese, D. M. 1958. Studies of the basaltic soils of northern Ireland. VI. Cation exchange capacities of the fine sand separates. *J. Soil Sci.* 9:289–297.

McCown, R. L. 1971. Available water storage in a range of soils in northeastern Queensland. *Aust. J. Exp. Agric. Anim. Husb.* 11:343–348.

McIntyre, D. S. and C. B. Tanner. 1959. A normally distributed soil physical measurement and non-parametric statistic. *Soil Sci.* 88:133–137.

Meybeck, M. 1983. Atmospheric inputs and river transport of dissolved substances. *IAHS Publ.* 141:173–192.

Møberg, J. P., I. E. Esu, and W. B. Malgwi. 1991. Characteristics and constituent composition of Harmattan dust falling in northern Nigeria. *Geoderma* 48:73–81.

Nicholls, P. H., R. H. Bromilow, and T. M. Addiscott. 1982. Measured and simulated behaviour of Fluometuron, Aldoxycarb and chloride ion in a fallow structured soil. *Pest. Sci.* 13:475–483.

Nielsen, D. R. and J. W. Biggar. 1962. Miscible displacement. 3. Theoretical considerations. *Soil Sci. Soc. Am. Proc.* 26:216–221.

Nielsen, D. R., J. W. Biggar, and K. T. Erh. 1973. Spatial variability of field-measured soil-water properties. *Hilgardia* 42:215–260.

Nye, P. H. and P. B. Tinker. 1977. Solute movement in the soil-root system. *Stud. Ecol.* 4.

Otoma, S. and T. Kuboi. 1988. Prediction of time variation in water and chloride profiles in a soil subject to cropping and annual application of sewage sludge. *J. Hydrol.* 99:1–17.

Pandey, R. N. and S. K. Gupta. 1978. Equations to predict leaching of soluble salts in saline soils. *J. Agric. Sci.* 91:131–133.

Panin, P. I. 1962. Salt discharge from soils and determination of leaching norms. *Soviet Soil Sci.* 7:703–709.

Passioura, J. B. 1971. Hydrodynamic dispersion in aggregated media. *Soil Sci.* 111:339–344.

Passioura, J. B. and D. A. Rose. 1971. Hydrodynamic dispersion in aggregated media. 2. Effects of velocity and aggregate size. *Soil Sci.* 111:345–351.

Peck, A. J. and D. H. Hurle. 1973. Chloride balance of some farmed and forested catchments. *Water Resour. Res.* 9:648–657

Peck, A. J., C. D. Johnston, and D. R. Williamson. 1981. Analyses of solute distributions in deeply weathered soils. *Agric. Water Manage.* 4:83–102.

Probert, M. E. 1976. The composition of rainwater at two sites near Townsville, Queensland. *Aust. J. Soil Res.* 14:397–402.

Raats, P. A. C. 1978a. Convective transport of solutes by steady flows. I. General theory. *Agric. Water Manage.* 1:201–218.

Raats, P. A. C. 1978b. Convective transport of solutes by steady flows. II. Specific flow problems. *Agric. Water Manage.* 1:219–232.

Reynolds, B. and A. B. Pomeroy. 1988. Hydrogeochemistry of chloride in an upland catchment in mid-Wales. *J. Hydrol.* 99:19–32.

Rhoades, J. D., D. B. Kreuger, and M. J. Reed. 1968. The effects of soil-mineral weathering on the sodium hazard of irrigation waters. *Soil Sci. Soc. Am. Proc.* 32:643–647.

Robbins, C. W., R. J. Wagenet, and J. J. Jurinak. 1980. A combined salt transport-chemical equilibrium model for calcareous and gypsiferous soils. *Soil Sci. Soc. Am. J.* 44:1191–1194.

Rogowski, A. S. 1972. Watershed physics: soil variability criteria. *Water Resour. Res.* 8:1015–1023.

Rose, C. W., P. W. A. Dayananda, D. R. Nielsen, and J. M. Biggar. 1979. Long-term solute dynamics and hydrology in irrigated slowly permeable soils. *Irrig. Sci.* 1:77–87.

Rose, C. W., F. W. Chichester, J. R. Williams, and J. T. Ritchie. 1982a. A contribution to simplified models of field solute transport. *J. Environ. Qual.* 11:146–150.

Rose, C. W., F. W. Chichester, J. R. Williams, and J. T. Ritchie. 1982b. Application of an approximate analytic method of computing solute profiles with dispersion in soils. *J. Environ. Qual.* 11:151–155.

Rose, C. W., W. L. Hogarth, and P. W. A. Dayananda. 1982c. Movement of peak solute concentration position by leaching in a non-sorbing soil. *Aust. J. Soil Res.* 20:23–36.

Rose, C. W., F. W. Chichester, and I. Phillips. 1983. Nitrogen-15-labelled nitrate transport in a soil with fissured shale substream. *J. Environ. Qual.* 12:249–252.

Rose, D. A. 1977. Hydrodynamic dispersion in porous materials. *Soil Sci.* 123:277–283.

Rose, D. A. and J. B. Passioura. 1971. The analysis of experiments on hydrodynamic dispersion. *Soil Sci.* 111:252–257.

Rowell, D. L., M. W. Martin, and P. H. Nye. 1967. The measurement and mechanism of ion diffusion in soils. III. The effect of moisture content and soil solution concentration on the self-diffusion of ions in soils. *J. Soil Sci.* 18:204–222.

Ruprecht, J. K. and N. J. Schofield. 1989. Analysis of streamflow generation following deforestation in south west Western Australia. *J. Hydrol.* 105:45–62.

Ruprecht, J. K. and N. J. Schofield. 1990. Streamflow generation following agricultural clearing. *Land Water Res. News* 2:19–20.

Russell, J. S., A. W. Moore, and J. E. Coaldrake. 1967. Relationships between subtropical, semiarid forest of *Acacia harpophylla* (Brigalow), microrelief, and chemical properties of associated gilgai soil. *Aust. J. Bot.* 15:481–498.

Russo, D. 1984. A geostatistical approach to solute transport in heterogeneous fields and its applications to salinity management. *Water Resour. Res.* 20:1260–1270.

Russo, D. and E. Bresler. 1981. Effect of field variability in soil hydraulic properties on solutions of unsaturated water and salt flows. *Soil Sci. Soc. Am. J.* 45:675–681.

Schofield, N. J. 1988. Predicting the effects of land disturbances on stream salinity in south-west Western Australia. *Aust. J. Soil Res.* 26:425–438.

Schofield, N. J., J. K. Ruprecht, and I. C. Loh. 1988. The Impact of Agricultural Development on the Salinity of Surface Water Resources of South-West Western Australia. Report No. WS 27. Water Authority of Western Australia.

Scofield, C. S. 1924. The movement of water in irrigated soils. *J. Agric. Res.* 27:617–693.

Scotter, D. R., B. E. Clothier, and T. J. Sauer. 1988. A critical assessment of the role of measured hydraulic properties in the simulation of adsorption, infiltration and redistribution of soil water. *Agric. Water Manage.* 15:73–86.

Sears, S. O. and D. Langmuir. 1982. Sorption and mineral equilibria controls on moisture chemistry in a C-horizon soil. *J. Hydrol.* 56:287–308.

Selim, H. M., J. M. Davidson, and P. S. C. Rao. 1977. Transport of reactive solutes through multilayered soils. *Soil Sci. Soc. Am. J.* 41:3–10.

Seyfried, M. S. and P. S. C. Rao. 1987. Solute transport in undisturbed columns of an aggregated tropical soil: preferential flow effects. *Soil Sci. Soc. Am. J.* 51:1434–1444.

Shalhevet, J. and L. Bernstein. 1968. Effects of vertically heterogeneous soil salinity on plant growth and water uptake. *Soil Sci.* 106:85–93.

Shalhevet, J. and J. Kamburov. 1976. Irrigation and Salinity, a World Wide Survey. International Commission on Irrigation and Drainage, New Delhi, India.

Sharma, M. L. 1988. Recharge estimation from the depth-distribution of environmental chloride in the unsaturated zone — Western Australia. In: Estimation of Natural Groundwater Recharge. I. Simmers (Ed.). NATO ASI Ser. C, Math. Phys. Sci., 222:159–173.

Sharma, M. L., P. Farrington, and M. S. Fernie. 1983. Localised groundwater recharge on the "Gnangara Mound," Western Australia. In: Proceedings of the International Conference on Groundwater and Man, Sydney, Australia, pp. 293–302.

Sharma, M. L., R. J. W. Barron, and M. S. Fernie. 1987. Areal distribution of infiltration parameters and some soil physical properties in lateritic catchments. *J. Hydrol.* 94:109–127.

Shaw, R. J. 1988a. Soil salinity and sodicity. In: Understanding Soils and Soil Data. I. F. Fergus (Ed.). Australian Society of Soil Science, Queensland Branch, Brisbane, Australia, pp. 109–134.

Shaw, R. J. 1988b. Predicting deep drainage in the soil from soil properties and rainfall. *Soil Use Manage.* 4:120–123.

Shaw, R. J. and P. J. Thorburn. 1985a. Prediction of leaching fraction from soil properties, irrigation water and rainfall. *Irrig. Sci.* 6:73–83.

Shaw, R. J. and P. J. Thorburn. 1985b. Towards a quantitative assessment of water quality for irrigation. In: Proc. 5th Afro-Asian Regional Conference. International Commission on Irrigation and Drainage, Townsville, Queensland, Australia.

Shaw, R. J. and D. F. Yule. 1978. Assessment of Soils for Irrigation, Emerald, Queensland. Agricultural Chemistry Branch Technical Report No. 13. Queensland Department of Primary Industries, Queensland, Australia.

Shaw, R. J., K. K. Hughes, P. J. Thorburn, and A. J. Dowling. 1987. Landscape, Soil and Water Salinity. A. Principles of Landscape, Soil and Water Salinity — Processes and Management Options. Proc. Brisbane Regional Salinity Workshop, May, Brisbane, Australia. Queensland Department of Primary Industries Conference and Workshop Series QC87003.

Simmons, C. S. 1982. A stochastic-convective transport representation of dispersion in one-dimensional porous media systems. *Water Resour. Res.* 18:1193–1214.

Slavich, P. G. and J. Yang. 1990. Estimation of field scale leaching rates from chloride mass balance and electromagnetic induction measurements. *Irrig. Sci.* 11:7–14.

Smiles, D. E., J. R. Philip, J. H. Knight, and D. E. Elrick. 1978. Hydrodynamic dispersion during absorption of water by soil. *Soil Sci. Soc. Am. J.* 42:229–234.

Smith, R. J. and N. H. Hancock. 1986. Leaching requirement of irrigated soils. *Agric. Water Manage.* 11:13–22.

Smith, S. J., L. R. Ahuja, and J. D. Ross. 1984. Leaching of a soluble chemical under field crop conditions. *Soil Sci. Am. J.* 48:252–258.

Snowy Mountains Engineering Corporation (SMEC). 1988. The Importance of Surface Water Resources Data to Australia. Water Management Series No. 16. Australian Water Resources Council.

Soil Survey Staff. 1975. Soil Taxonomy. A Basic System of Soil Classification for Making and Interpreting Soil Surveys. Agricultural Handbook No. 436. U.S. Department of Agriculture, U.S. Government Printing Office, Washington, D.C.

Starr, J. L., J. Y. Parlange, and C. R. Frink. 1986. Water and chloride movement through a layered field soil. *Soil Sci. Soc. Am. J.* 50:1384–1390.

Steenhuis, T. S., T. L. Richard, M. B. Parlange, S. O. Aburime, L. D. Geohring, and J. Y. Parlange. 1988. Preferential flow influences on drainage of shallow sloping soils. *Agric. Water Manage.* 14:137–151.

Stewart, J. B. 1984. Measurement and prediction of evaporation from forested and agricultural catchments. *Agric. Water Manage.* 8:1–28.

Suarez, D. L. 1986. A soil water extractor that minimises CO_2 degassing and pH errors. *Water Resour. Res.* 22:876–880.

Sukhija, B. S., D. V. Reddy, P. Nagabhushanam, and R. Chand. 1988. Validity of the environmental chloride method for recharge evaluation of coastal aquifers, India. *J. Hydrol.* 99:349–366.

Szabolcs, I. 1986. Agronomical and ecological impact of irrigation an soil and water salinity. *Adv. Soil Sci.* 4:189–218.

Talbot, R. J. and I. J. Bruce. 1974. Effects of leaching on salinity levels and other soil properties of a clay soil in the Lockyer Valley, Queensland. *Queensland J. Agric. Animal Sci.* 31:175–183.

Thomas, G. W. and N. T. Coleman. 1959. A chromatographic approach to the leaching of fertiliser salts in soils. *Soil Sci. Soc. Am. Proc.* 23:113–116.

Thomas, G. W. and R. E. Phillips. 1979. Consequences of water movement in macropores. *J. Environ. Qual.* 8:149–152.

Thorburn, P. J. and C. W. Rose. 1990. Interpretation of solute profile dynamics in irrigated soils. III. A simple model of bypass flow in soils. *Irrig. Sci.* 11:219–225.

Thorburn, P. J., C. W. Rose, and D. F. Yule. 1990a. Interpreting solute profile dynamics in irrigated soils. II. Convective-dispersive approach. *Irrig. Sci.* 11:209–217.

Thorburn, P. J., C. W. Rose, R. J. Shaw, and D. F. Yule. 1990b. Interpretation of solute profile dynamics in irrigated soils. I. Mass balance approaches. *Irrig. Sci.* 11:199–207.

Thorburn P. J., B. A. Cowie, and P. A. Lawrence. 1991. Effect of land development on groundwater recharge determined from non-steady chloride profiles. *J. Hydrol.* 124:43–58.

Tillotson, W. R. and R. J. Wagenet. 1982. Simulation of fertiliser nitrogen under cropped situations. *Soil Sci.* 133:133–143.

Trudgill, S. T., H. A. Viles, R. J. Inkpen, and R. U. Cooke. 1989. Remeasurement of weathering rates, St Paul's Cathedral, London. *Earth Surface Proc. Landforms* 14:175–196.

U.S. Bureau of Reclamation (USBR). 1982. Land Classification—Techniques and Standards. Part 513.1.3c(2)(d). Reclamation Instructions Series 510. U.S. Bureau of Reclamation, Washington, D.C.

U.S. Salinity Laboratory (USSL). 1954. Diagnosis and Improvement of Saline and Alkali Soils. Agricultural Handbook No. 60. U.S. Department of Agriculture. U.S. Government Printing Office, Washington, D.C.

Van de Pol, R. M., P. J. Wierenga, and D. R. Nielsen. 1977. Solute movement in a field soil. *Soil Sci. Soc. Am. J.* 41:10–13.

Van der Molen, W. H. 1956. Desalinisation of saline soils as a column process. *Soil Sci.* 81:19–27.

Van Genuchten, M. Th. and W. J. Alves. 1982. Analytical solutions of the convective-dispersive solute transport equation. Technical Bulletin No. 1601. U.S. Department of Agriculture, Washington, D.C.

Van Genuchten, M. Th. and F. N. Dalton. 1986. Models for simulating salt movement in aggregated field soils. *Geoderma* 38:165–183.

Velbel, M. A. 1989. Discussion of "Rates of soil formation: implications for soil-loss tolerance," by E. B. Alexander. *Soil Sci.* 148:71–74.

Wagenet, R. J. 1986. Water and solute flux. In: Methods of Soil Analysis, Part 1. Physical and Mineralogical Methods. Agronomy Monograph No. 9 (2nd ed.). American Society of Agronomy-Soil Science Society of America, Madison, WI, pp. 1055–1088.

Wagenet, R. J. and J. L. Hutson. 1984. LEACHM: A process based model of water and solute movement, transformations, plant uptake and chemical reactions in the unsaturated zone. Continuum 2. Publication of the Water Resources Institute, Center for Environmental Research, Cornell University, Ithaca, NY p. 80.

Wagenet, R. J. and B. K. Rao. 1983. Description of nitrogen movement in the presence of spatially-variable soil hydraulic properties. *Agric. Water Manage.* 6:227–242.

Walker, G. R., I. D. Jolly, and P. G. Cook. 1991. A new leaching approach to the estimation of diffuse recharge following a change in land use. *J. Hydrol.*128:49–67.

Walling, D. E. and B. W. Webb. 1983. The dissolved loads of rivers: a global overview. In: Dissolved Loads of Rivers and Surface Water Quantity/Quality Relationships. IAHS Publ. No. 141. Proc. Hamburg Symposium, August, 1983. pp. 3–20.

Walling, D. E. and B. W. Webb. 1986. Solutes in river systems. In: Solute Processes. S. T. Trudgill (Ed.). John Wiley & Sons, New York.

Warington, R. 1900. Lectures on Some of the Physical Properties of Soil. Clarendon Press, Oxford.

Warrick, A. W., J. W. Biggar, and D. R. Nielsen. 1971. Simultaneous solute and water transfer for an unsaturated soil. *Water Resour. Res.* 7:1216–1225.

Warrick, A. W., J. H. Kichen, and J. L. Thames. 1972. Solutions for miscible displacement of soil water with time-dependent velocity and dispersion co-efficients. *Soil Sci. Soc. Am. Proc.* 36:863–867.

Watson, K. W. and R. J. Luxmoore. 1986. Estimating macroporosity in a forest watershed by use of a tension infiltrometer. *Soil Sci. Soc. Am. J.* 50:578–582

White, R. E. 1985a. The transport of chloride and non-diffusible solutes through soil. *Irrig. Sci.* 6:3–10.

White, R. E. 1985b. The analysis of solute breakthrough curves to predict water redistribution during unsteady flow through undisturbed structured clay soil. *J. Hydrol.* 79:21–35.

White, R. E. 1985c. The influence of macropores on the transport of dissolved and suspended matter through soil. *Adv. Soil Sci.* 3:95–120.

White, R. E., S. R. Wellings, and J. P. Bell. 1983. Seasonal variations in nitrate leaching in structured clay soils under mixed land use. *Agric. Water Manage.* 7:391–410.

White, R. E., G. W. Thomas, and M. S. Smith. 1984. Modelling water flow through undisturbed soil cores using a transfer function model derived from ^{3}HOH and Cl transport. *J. Soil Sci.* 35:159–168.

White, R. E., J. S. Dyson, Z. Gerstl, and B. Yaron. 1986a. Leaching of herbicides through undisturbed cores of a structured clay soil. *Soil Sci. Soc. Am. J.* 50:277–283.

White, R. E., J. S. Dyson, R. A. Haigh, W. A. Jury, and G. Sposito. 1986b. A transfer function model of solute transport through soil. 2. Illustrative applications. *Water Resour. Res.* 22:248–254.

Wierenga, P. J. 1977. Solute distribution profile computed with steady state and transient water movement models. *Soil Sci. Soc. Am. J.* 41: 1050–1055.

Wierenga, P. J. and M. Th. van Genutchen. 1989. Solute transport through small and large unsaturated soil columns. *Groundwater* 27:35–42.

Willardson, L. S. and R. J. Hanks. 1976. Irrigation management affecting quality and quantity of return flow. Report EPA-600/2-76-226. Environmental Protection Technology Series. U.S. Environmental Protection Agency.

Wright, C. H. 1928. The specific conductivities of soil extracts. *J. Agric. Sci.* 18:186–193.

Yaalon, D. H. 1983. Climate, time and soil development. In: Pedogenesis and Soil Taxonomy. I. Concepts and Interactions. L. P. Wilding, N. E. Smeck, and G. F. Hall (Eds.). Elsevier, Amsterdam, pp. 233–251.

Yaalon, D. H., D. Mor, and M. Netzar. 1984. Basaltic weathering rates and associated soils on the Golan Heights, Israel. Unpublished abstract.

Yule, D. F. and W. R. Gardner. 1978. Longitudinal and transverse dispersion co-efficients in unsaturated plainfield sand. *Water Resour. Res.* 14:528–588.

Zein el Abedine, A., G. H. Robinson, and J. Tyego. 1969. A study of certain physical properties of a vertisol in the Gezira area, Republic of Sudan. *Soil Sci.* 108:359–366.

Zimmermann, U., K.-O. Münich, and W. Roether. 1967. Downward movement of soil moisture traced by means of hydrogen isotopes. Geophysical Monograph No. 11. American Geophysical Union, pp. 28–35.

Index